D0458327

Wood Heat

Other Books by the Same Author

The Manual of Practical Homesteading
Building Stone Walls

Wood Heat

BY

John Vivian

ILLUSTRATED BY LIZ BUELL

Rodale Press, Inc.
Emmaus, Pennsylvania

PRINTED IN THE UNITED STATES OF AMERICA
on recycled paper

ISBN 0-87857-131-0 hardcover

ISBN 0-87857-149-3 paperback

Library of Congress Cataloging in Publication Data

Vivian, John
Wood heat.
Bibliography: p.
Includes index.
1. Stoves, Wood. 2. Fireplaces. 3. Wood as fuel.
4. Heating. I. Title.
TH7437.V58 697'.04 76-28539
ISBN 0-87857-131-0
ISBN 0-87857-149-3 pbk.

4 6 8 10 9 7 5

For Susa

Contents

"I smell yo' bread a-burnin'; turn yo' damper down.
If you ain't got a damper, good gal, turn yo' bread around."

Jimmy Rodgers
"Mule Skinner Blues"
(Blue Yodel Number 8)

BISCUIT
BUSTER

Introduction

Almost a decade and a half have gone by now since Louise and I first decided to search out the hard but rewarding life or rural or semirural self-sufficiency that has come to be known as modern homesteading. One of our earliest goals was to heat and cook with wood, originally out of a native but harmless romance with doing things the good old-time way, I guess. Plus a determination to become free of the giant oil companies—borne of nothing but an ornery independence. It was years before the energy crisis of the mid-1970s would force fuel prices sky-high. Only a farsighted few (we not among them at the time, I must admit) realized how limited was our supply of petroleum and natural gas; how the sulphur in coal posed a threat to air quality while the strip-mining techniques used to get it out endangered the land; and how nuclear power was to create at least as many problems as it cured. Why, in those days some folks looked down their noses at wood burners, thinking us way behind the times without a modern gas or oil furnace. And here just a few years later we are all coming to realize that wood is just about *the most* timely and up-to-date energy source there is. Wood alone is self-perpetuating, the only truly renewable fuel source available in these increasingly resource-short times.

Like most moderns, Louise and I embarked on our wood heating experience with no more than a typical city person's knowledge of wood fires—how to start a camp fire (by rubbing two Boy Scouts together, my mother would say) and how to keep a cheery party fire going in the gas-augmented recreation room fireplace. Over several years we gained the know-how and financial stability to make a complete break with the conventional economy. We also managed to pick up a few more bits of wood heat skill. It took several burned fingers, but I coaxed the fireplace in our city apartment to roast a fair leg of lamb—that is if we weren't too picky about the degree of over- or underdoneness.

On our first country place, a recreational farm within commuting distance of the city, we managed to keep things warm (if smoky) on weekends with fires in the walk-in fireplace that came with the old Pennsylvania Dutch stone house. Louise learned the most there, spending hours patiently moving the Dutch oven hanging from a big cooking crane around and shovelling coals here and there to achieve proper cooking heats. She became a fair open-hearth cook, to the point even of using a big Dutch oven to bake yeast breads that came out delectable four times in five. I kept up my own cookery, becoming skilled enough that the foil-wrapped sweet corn and potatoes from the garden were only burned on one side instead of halfway through when roasted in coals under the spitted roast.

But this was for fun and learning. The wood fires were mainly entertainment, their practical benefits secondary. We really knew nothing about serious wood heat, a fact I proved shortly after we packed up the decade-old Ford pickup and farm trailer with our belongings and headed off to an old house on a lonely Massachusetts hillside, to spend the next eight years living on the land for real.

Once at the farm, one of our first additions was a small Franklin-style open-front wood stove. It was a two-door type with a single cook lid on top, patented almost 200 years ago, and we were able to pick it up for just a few dollars back then. We'd found a stovepipe hole already chipped into the antique brick when we opened a flue that must have been plastered up since the Civil War era. Fortunately (it turned out) the hole was several inches higher than the stove's smoke outlet. To raise it I built a hearth from what was handy, old firebrick taken from a pile left when another chimney was torn down years ago.

Assuming that the stove would act like fireplaces we'd known, I connected it to the chimney with a short length of stovepipe. Louise put a kettle of cider with cloves and a few cinnamon sticks on top to mull properly. Then we scrounged around the barn for firewood. After packing the stove with dry pine limbs and dusty old maple chunks we lit a few newspapers under the grate and sat back to enjoy the first fire in our new home. That fire nearly took our home with it!

It was lovely at first as the dry pine caught and the bright flames quickly licked higher and higher, filling the evening dark with cheery warmth. The pine limbs did seem to burn unusually fast, but we just tossed on the rest of the old maple chunks and went out to the barn for more. But when we got back it was obvious that something was wrong. Even out in the entry we smelled an odd odor coming from the house—not the good smoke of a wood fire but a hot, metallic scent more like an overheated foundry. We rushed into the living room to find the fire going with a deep roar. The bed of pitch-filled pine coals was superheating the tinder-dry maple and the fire was becoming a growing inferno. White hot flame filled the firebox, pouring up and back into the chimney. The length of stovepipe showed a brightening glow, and the cider suddenly erupted into a rolling boil.

With our only fire tool, an undersized pair of tongs, I scrabbled at the stove doors, dodging spattering cider and popping sparks. The stovepipe was almost a cherry red now and the sides of the iron stove itself were beginning to show a dull crimson. I managed to get the doors pulled out and lodged partly closed, but the howling only grew louder and the fire hotter as air was pulled at greater speed through the restricted opening. The cider kettle began to thump on the stove top, and the stove itself, I swear, started to hum and shake as it poured out increasingly intense heat. I want you to know that I was shaking as I realized that our farm; indeed, our whole new life—just might become nothing but a pile of charred beams in a cellar hole.

Luckily we had picked up a garden hose at the same place we got the stove, and Louise rushed out after it when my attempt to close the stove doors appeared useless. But then, in she rushed, white-faced and more scared looking that I'd ever seen her, and beckoning me out. The night sky appeared alight! Our chimney was belching sparks and little licks of flame and cherry red coals were scattering over the roof. *And* the hose proved to be too short by half to reach into the house.

With her usual presence of mind Louise began to hose down the chimney and roof while I went back in and, unwittingly, made my only sensible move of the evening. With arms and hands wrapped in wet towels against the heat, I slowly poured the hot cider into the howling flames. Whereas cold water in any quantity could have caused the hot iron to split with near explosive force, the boiling liquid turned to steam on hitting the metal and the draft pulled the fire-quenching vapor into the coals and flames. Slowly the flame subsided and coals dimmed, and finally there was nothing but a steaming bed of black char.

The stove continued to radiate heat, but both the cast iron and sheet metal of the stovepipe quickly lost color, crackling as they contracted in cooling. When I stopped shaking enough to make it outside, the chimney blaze was out also—it turned out to be nothing but flame from last years' chimney swallow nests plastered on partway down the flue, rather than the truly serious tar-based type chimney fire that has brought down many an old house. We had a proper mess of burned cider and ashes inside, the stove never regained its color along the sides, (it looks bleached and ashy after a fire or two no matter how much stove black we put on), and the roof shingles of the house sport a scattering of burnt holes. Thanks mainly to the purely accidental brick hearth, which prevented the stove from burning its way through the pine-board floor and erupting in the cellar, our home was safe.

So, through dumb luck and a smiling Providence our homestead survived my ignorance of wood stoves. You may be sure that after that close call we set out quickly to learn about dampers (I hadn't installed one), draft controls and other fire regulators as well as cleaning flues, putting in heat shields, how to judge which wood to use in what quantities in which stage of dryness and at what stage in a fire's life. Because a fired wood stove is very much a living thing, one that must be attended carefully—or it just may kill you.

Most of this book is based on our own less harrowing experiences with wood heat. Wood was our primary fuel during our near-decade on a farm in the woods. When we traded most aspects of farm life but the woodlot and garden for a place closer to town over the duration of the kids' school years we learned how wood can be a town dweller's supplementary fuel—and reduce the gas or oil bill by three-quarters, or more. But wood heat is a broad subject. We don't pretend to have done it all and we welcome expert help. Liz Buell who did the illustrations assisted with more than the pictures. She and David and the kids live way up in North Sullivan, Maine, entirely on wood heat. They offered many good tips. We'll introduce other folks who helped out too, as we go along.

To start, let's go into the basic chemistry of wood heat, then the history of man's progression from pre-industrial dependence on wood for fuel, through the heyday of coal, then petroleum and now, for many of us it's back to wood. Next we will go into flues and chimneys. (A flue is the long, vertical opening inside the chimney. You might say that a chimney is what surrounds a flue.) Flues aren't terribly exciting, but are the most important part of a wood fire and we'll cover all kinds, permanent types and intermittant-use flues of stovepipe or "cattied" of sticks and mud that you can build yourself. Permanent flues of brick or cement block should be erected by a mason, but we'll show how you can inspect and repair your own when feasible. I will also give directions on building a cement block chimney if you have the ambition.

Wood stoves for heat are next and we'll discuss the pros and cons of the major types available—antiques, Scandinavian and other European imports, U.S. and Canadian-made. We'll do our best to help you evaluate old stoves, then put them back into working condition. There are photos of restoring an old heater. We'll also picture the job of making a stove from a steel drum, and will install several kinds of stove, in different typical locations, then get them fired up right. We'll end up with central heat from wood or a wood/oil or wood/gas furnace.

Fireplaces are not as efficient heaters as stoves are, but a lot of homes have them. So we'll try to help you make yours as efficient as possible. I'll admit a bias in favor of a fireplace design you may have heard of—the "Rumford," devised by a contemporary of Ben Franklin (who in turn designed the Franklin fireplace stove I'm sure you've heard of). Unlike Ben Franklin, Rumford took the wrong side in the American Revolution, he doesn't get fair treatment in our history books. But his fireplace is a dandy and we'll describe it in enough detail that you can put one in if you're building a new house or extensively remodeling an old one.

Louise is our cooking expert and the following chapter will include all she has to tell about cooking on the open hearth, stoves and wood ranges. Plus a lot of information provided by Liz

Buell and other wood cookers. (Each wood stove—the oven in a wood range in particular—is an individual. No wood cooker can tell you precisely how to operate yours, but their combined experience with their own can get you started.) The paraphernalia associated with wood cooking is fun: trammels and trivets, danglespits and gridirons. And the food is better than anything a modern gas or electric stove ever turned out.

Then we come to wood itself. Many people will be purchasing their supply from a commercial woodcutter, so we'll try to help find you a good one and judge his wood. For those, like us, who prefer to cut as much of their own wood as they can, we'll touch on finding and evaluating a woodlot, managing it for fuel, and the basics of felling trees, aging and getting in the wood.

The fringe benefits of wood heat come last. What ashes you don't use to make soap with are great plant food and protection in the garden. So is the soot from stovepipe or chimney cleanout. And the texture that a good-heating fireplace or stove gives to life is something that is hard to put into words. But, I'll tell you that I miss that friendly, warmth-radiating part of the home if the fire goes out. So will you.

Then, to end up there is a bit of speculation about the future of wood heat and a list of suppliers of equipment and a bibliography of books and articles that's as complete as we could make it, so you can check out all the available information on this old but just rediscovered way of life. And that's what wood heat is, really. More than a way to keep warm, it's an expression of a more natural, conserving attitude toward this threatened globe and all who inhabit it. There's even a political side to wood heat and we'll get into that when the opportunity arises. So, let's get going!

CHAPTER ONE
The Science and History of Wood Heat

Before we get into the details of how to heat and cook with wood, let's check back briefly with the chemistry books for an understanding of the way wood is produced and what happens when you burn it. If we take the time to gain a basic understanding of the "why" of the whole process, the "how" will make a lot more sense.

Wood Heat Chemistry

It is a basic fact of existence on our planet that (nuclear power aside) every smidgin of useful energy at our disposal originated as rays from the sun stored up in winds, rain clouds, tides, or in the case of fossil fuels and wood, by green chlorophyll-bearing plants. The process, called photosynthesis, is as close to magic as we are likely to see. Somehow, (scientists can't duplicate it artificially in the laboratory) the chlorophyll in plants' green leaves and stems takes the elements contained in two inert compounds, water and carbon dioxide, and uses the energy of sunlight to turn them into a carbohydrate, glucose sugar. The elements are carbon, hydrogen and oxygen. The chemical formula, if you are interested, is $6(H_2O) + 6(CO_2) + \text{ENERGY} = C_6H_{12}O_6 + 6(O_2)$.

Now this deceptively simple process is the basis of all life as we know it; the plant sugars are the building blocks of all we and other living things eat. In addition, the six molecules of O_2 the plants put out along with the sugar provide the oxygen that you and I breathe. But what interests us here is not the food and air that plants offer us, but the energy they pack away while doing it.

Look at that formula again, the energy part you see on the left side of the = sign doesn't appear on the right side. That energy is trapped in the glucose sugar. Now, as they grow, many

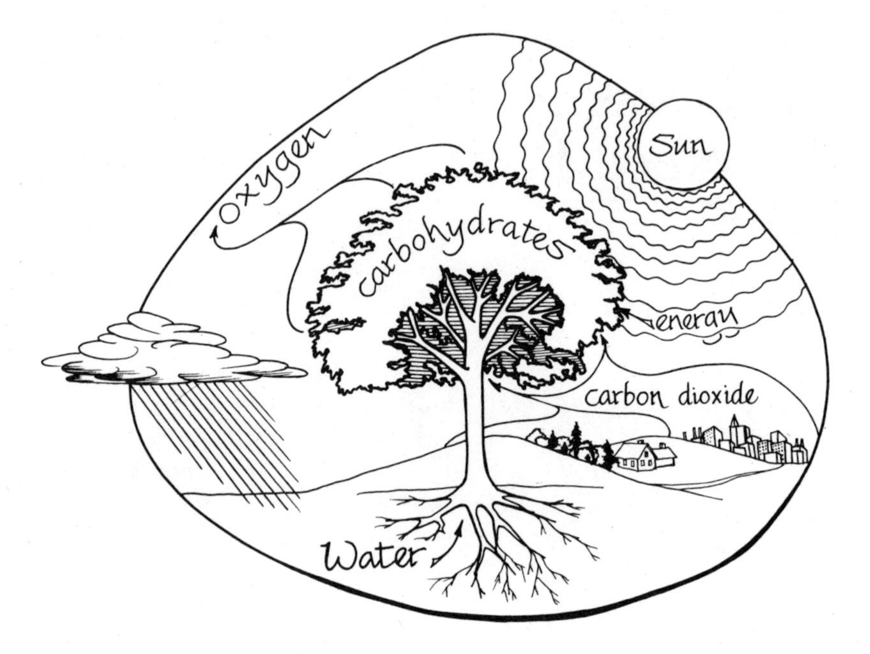

plants — most trees — take some of the sugar they manufacture and turn it into cellulose (glucose sugar is $C_6 H_{12} O_6$ while cellulose is $C_6 H_{10} O_5$). Not much apparent difference on paper, but by removing a little bit of hydrogen and oxygen the plant turns sugar into one of the most indestructible compounds. Cellulose is what makes wood hard. Chopped, cooked and spread thin, cellulose fiber becomes paper—and, so tough is it that the fiber can be reused many times. Indeed, the book you are reading now is little more than cellulose fiber—recycled. It's been used at least once before, at that.

Now, there are relatively few things in nature that can harm cellulose. The only creatures that can digest it are a few specialized types of bacteria—including one kind that lives in the innards of termites. The termite feeds the bacteria, and they feed the bug. But even with the help of termites, natural decomposition of a mature hardwood tree takes many years. About the only natural force that can destroy wood in a hurry is fire. Once you start a log burning it will continue on its own, in more or less a reverse of the photosynthesis process called combustion, or rapid oxidation.

During oxidation, free oxygen is added back into the chemical process, turning the wood back into its basic compo-

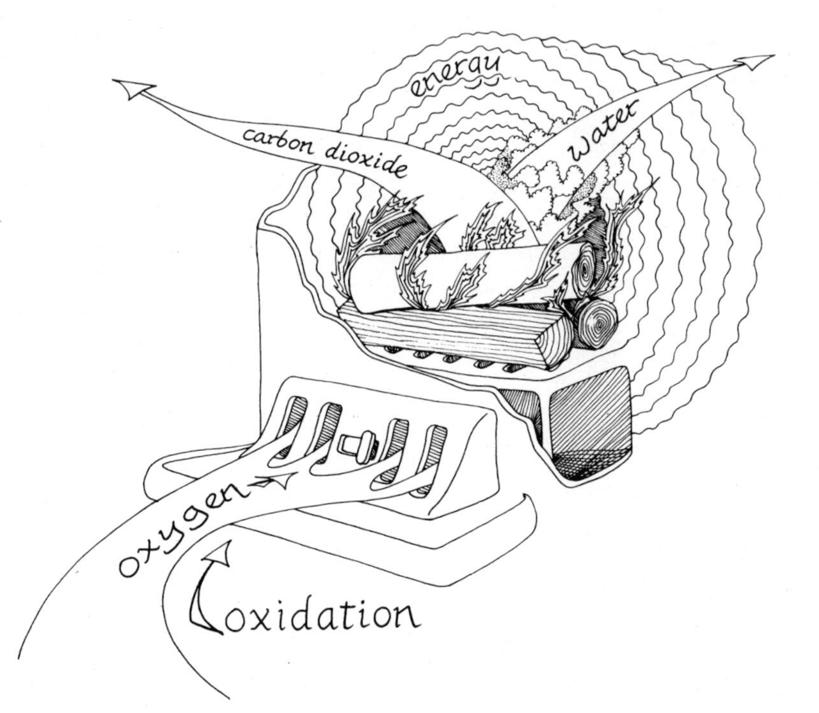

nents, CO_2 and water (plus a few other compounds we'll get into later). Most importantly, the sun energy that went into making the wood is now released as light and heat. Needless to say, the more oxygen you supply a fire, the faster the wood will oxidize—the faster, hotter and brighter it will burn. This is why you blow on a fire or put a bellows to it to get it going.

Fire Safety

So—chemically, a wood fire is nothing but a reversal of nature's growth process, but at greater speed. Oxygen supply is the key to the fire's intensity; if we can control the supply of oxygen the fire receives, we are in control of the fire itself. Wood stoves, fireplaces and flues come equipped with a battery of dampers, draft controls, doors and such that help control the air your fire receives. Learn to use them properly and you are on the way to having a safe fire. On the other half of safety, much has to do with proper choice of stove, correct placement of stove and pipe, construction and cleaning of flues. But, with any fire, there is always the chance of a mistake, a flawed pipe or fire screen that can let out an errant spark.

Just one spark can burn your house down and you with it. So before you light that first fire, get yourself a good fire extinguisher. If the experience that Louise and I had with our first blaze is any indication, an extinguisher should be a higher priority on your "buy list" than that first stove or set of andirons for the fireplace. And even if it is long enough, don't rely on your garden hose; it may be frozen up or disconnected, or the fire can put your pump wiring out of commission. Don't settle for one of the old-fashioned water or "dry" extinguishers. Get a pressurized, universal extinguisher and hang it so it is handy to your stove or fireplace.

Fire Extinguishers

The best (universal) extinguishers are rated to extinguish fires of A, B and C class—routine, grease and electrical fires. Plain water will do for most wood fires—Class A. But what if you are frying bacon on the stove when it overheats, then suddenly the bacon grease catches and before you can get ahead of the

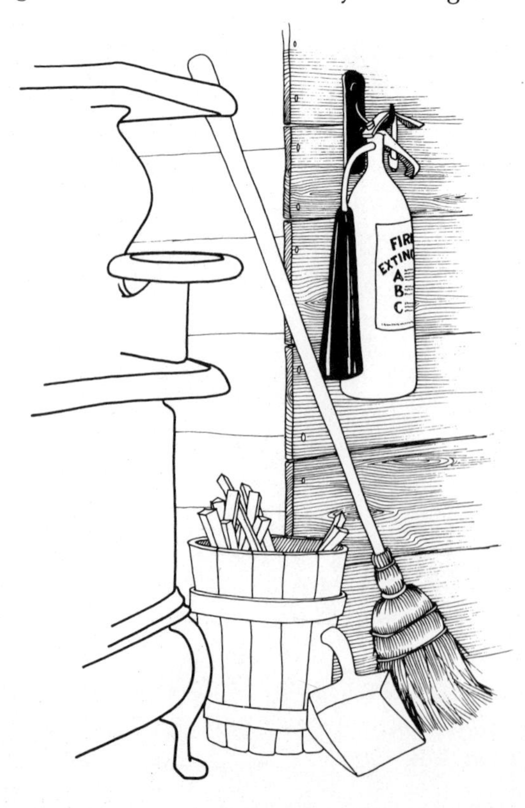

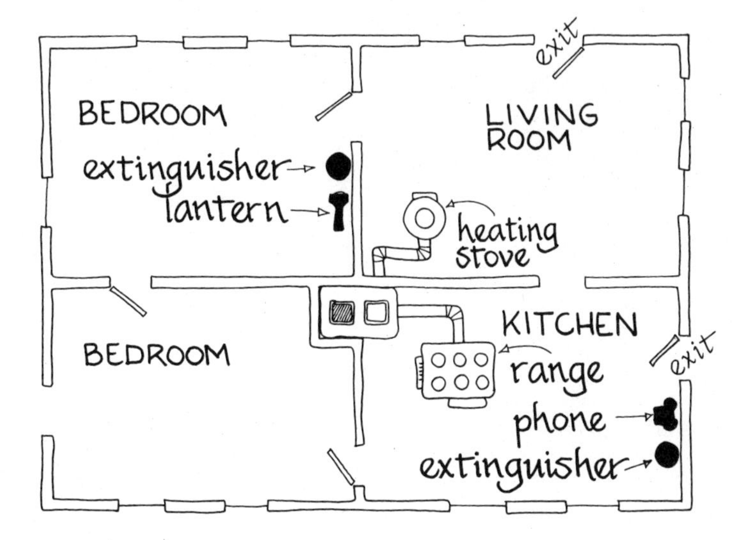

sputtering, smoking grease fire, it proceeds to fuse the wires in the light socket overhead? A highly unlikely set of coincidences, I'll agree. But it has happened. So, save yourself worry—and maybe save your home, to say nothing of some pretty important lives. Get the biggest, and best all-class extinguisher you can find. Better, get two. Put one on the wall between your fire and the nearest exit—and it's a good idea to have your telephone there too. That way, you can clear the house of people, call the fire department and extinguish the fire as you go in. Moreover, the exit is always at your rear, so even if you lose to the fire, you won't lose everything.

Put the other extinguisher somewhere between the stove or fireplace and your sleeping quarters. Many homes have bedrooms on a second story. Put an extinguisher in a handy place on a wall on the upstairs landing so you can grab and go without thinking about it if roused from a deep sleep by smoke. It's best to have a good, well-charged electric lantern in the same spot too. And, needless to say, have the family well rehearsed for a quick escape from the upper story or stories. On our place, we can get out on the porch roofs from any number of upstairs windows. Lacking a house so designed, you'd best also put in fire stairs, rope ladder or knotted ropes for fire escapes.

I'm not trying to scare you off wood heat with all these warnings—far from it. But modern home heating and cooking technology has just about removed the danger of appliance-caused home fires. Today's electrical stoves and heating units have circuit breakers, gas ranges have automatic cutoffs that stop gas flow when the fire goes out, and home gas and oil furnaces have a half-dozen failsafe mechanisms—all designed to protect you without any thought or effort on your own part. Well, fine and good and three cheers for the Home Heat Council.

But with wood heat of any sort, we are getting into an earlier technology, back to the appliances of an era when machines did less for people. Then, you split your own wood yourself, stoked your own fire and had the know-how to provide for your family's safety. Trouble is, even back in the all-wood-heated 1800s, houses burned down despite generations of built-up caution and knowledge. Today's neophyte wood burner, who like ourselves was probably raised around self-cleaning ovens and all-electric heat, takes on the chore in blissful ignorance of the potential dangers. Or at least Louise and I did, assuming without thinking that a wood stove was just as safe as the complicated oil furnaces we'd had in suburbia. (Of course, we were blissfully ignorant of the complex workings of the oil burner too—still are, for that matter. Something goes wrong, and you just call the serviceman, right?)

Well, there's no serviceman for the wood heater, and no need. In place of all the automatic gadgets that he has been trained to keep going, you'll have only a set of controls that appear simple, but that can be manipulated in an amazing number of ways, plus your own sense. Or your own senses, I should say. In time you'll be able to tell by the sounds your equipment makes, by the smells it lets off and the intensity of heat and cooking speed, just how things are going. And you should know; or be able to find out quickly—at any hour of day or night.

Caution

The secret of safe wood burning, I guess — after knowing precisely what you are doing — is to "heat 'n' sleep scared." You don't have to live terrified by any stretch of the imagination, and certainly shouldn't exist in fear borne of ignorance or uncertainty. But a healthy caution is part of everything about wood heat. In cutting the wood, you keep a sound respect for the angle of the ax stroke, so it won't skip out of the cut and put you on crutches. And, before the tree fall gets much underway, you hightail it; not only away from the falling trunk, but also to avoid the stump, that can swivel or split off the butt with great velocity. So it is with the wood fire; retain a knowledgeable respect for it, and it will heat and cook for you faithfully. I can recall many a night Louise or I have been awakened with a start by an odd sound or smell coming from the stoves.

To date, it's been nothing but a downdraft from a strong wind gust pushing smoke down the flue or a loud pop as a length of stovepipe cooled down or another harmless incident. That's so far—we continue to stay alert. With wood heat, you have to train your own sense and senses do what technology does for you in a modern home.

Oh yes, the modern electronic smoke detectors—the gadgets that can detect a whiff of smoke before you could begin to smell it, can be rigged up to call the fire department automatically. If I were you, I'd forget them. For one thing, if you rely on such a gadget you may feel safe enough that you won't let your own senses develop enough to detect such dangers as a flue fire where all the smoke goes up the chimney and the alarm won't sound till it's too late. And besides, with any wood-burning device you are bound to have agreeable smoke smells around when you start the blaze, lift a cover on the range to add more fuel or after a strong downdraft when someone opens the front door in winter—all of which could have the fire department suiting up and roaring to your door at odd hours. The alarms may be fine for modern smokeless houses, but for heating with wood, I'd say they are *too* sophisticated. So, for the last time, please use your own sense —common sense and the senses the Lord gave you to detect changes in your environment.

A History of Fuels in America

When the first British settlers reached what is now New England they were amazed to find the new land almost solidly wooded. The populated areas of their homeland, and of the Dutch lowlands in which many of them had spent several years of exile, had long since been cut over; most central European trees were, and still are, saved for their ornamental value or for building stock. Heat during the eighteenth century in those relatively mild climes was supplied by open fireplaces burning brush bundled into faggots, occasionally from hay or field grasses, and increasingly from peat or soft coal burned in stoves.

Wood, the First Fuel

But what a blessing all that wood was in the bitter winters of the colonists' new land. They built their homes of huge timbers and thick clapboards or shingles and packed the walls with insulating leaves. They heated with wood fires kept roaring in three, four and more open fireplaces arranged around a great chimney built in the center of the house. Designed on the European model, most heating fireplaces were relatively shallow, the fires built well out on the hearth so as to radiate as much heat as possible into the rooms. Often the kitchen fireplace was large enough to stand or sit in—and this the families often did on the coldest days.

But all flues were open, lacking dampers or any way of stopping the draft, short of stuffing cloth into the opening between fireplace and chimney. And, if there was not a good brisk fire in a fireplace, the draft quickly sucked the room clear of warmth. (One reason that rooms in old north country homes are small and have so many doors was to keep warm air in when the fireplace was going, and to keep the room and its drafty fireplace closed off when not in use.)

Back in Britain such chimney designs had been adequate to drive away the chill of the wet and cool winters. The brickwork of the flues would heat up, and together with a small but cheerful fire on the hearth would radiate enough warmth to keep the house comfortable. But not in New England with its subzero temperatures, howling nor'easters and month upon month of snow and ice. Here the fires had to blaze full force, and blaze they did—using tremendous amounts of wood. It is estimated that the original fireplaces were about ten percent efficient; that is, only one-tenth of the heating potential of the wood went to heat the houses and their occupants. The rest went up the chimney. But, the forests were full of trees, and the land had to be cleared anyway for crops. So, the original forests of New England literally went up in smoke, at the rate of four to five cords per person per season.

Farther south, in the Philadelphia area, many of the settlers were of German origin. In a somewhat less harsh climate they built of brick or stone with chimneys located at the ends of the houses. Indeed, the first farm that Louise and I lived on was in Pennsylvania Dutch country—with two-foot-thick stone walls and a huge walk-in fireplace. One of our neighbors lived in an even more traditional German-style farmhouse. The stable backed directly against the house, so animals and humans could share warmth (and, I presume, fragrances). At one side was a wood storage room with a fireplace built into the wall. Then inside the house was a large combination range and heater. The fire was fed from outside in the woodshed. Heat and smoke flowed through channels in the stove before entering the flue. This one was of brick, though in Europe they were often made of stone, some decorated with porcelain on brick or ceramic tiles. Of the later designs, many had doors attached onto the fire opening, permitting some regulation of air flow, thus affording a more efficient furnace.

Franklin's Fireplace

However, all the early wood-burning fireplaces were extravagantly wasteful of wood. Keeping a device such as our

Pennsylvania neighbors' fueled was near full-time work for a hired hand. As early as the mid-1700s shortages of fuel wood were cropping up here and there along the East Coast. One result was to stimulate the ever-inventive mind of Philadelphian Benjamin Franklin. He reasoned that the traditional fireplace permitted too much hot air to escape up the chimney. So, he developed one of the first dampers, a sliding door to restrict air flow between fire and flue. One of his concerns was that the closed rooms of his German-descended neighbors were unhealthily stuffy using the outside-fired furnaces. So, one of his most famous inventions, a freestanding fireplace, which evolved into today's "Franklin stove," had a cheerful British-style open fire to evacuate room air as the wood burned.

Franklin also observed that wood burned in long tongues, so that much of the flame itself was lost up a conventional chimney. The woodburning process, then as now, has three distinct phases. First, any water vapor must be turned to steam and boiled off before the wood can ignite. Second, the volatile oils—turpentine in pine trees, for example—must vaporize and burn. Being much lighter than air they rise quickly, producing the tongues of flame. Often oils plus water vapor will escape the firebed before they completely burn. This produces smoke, or as we'll see later the potential fire hazard, creosote. The final stage is combustion of the coals—nearly pure carbon—which burn very hot with little smoke or visible flame.

In his development, which he called the "Pennsylvanian fireplace," Franklin attempted to assure complete combustion by containing the flames and smoke within the firebox as long as possible. There was a false back in the box—a plate or hollow wall halfway back into the stove. The device was exhausted into an otherwise bricked-up fireplace. Flame and smoke had to pass up and around the center baffle, so the wood had more time to burn thoroughly. The cast-iron shell radiated air out into the room, and in some versions air was also drawn up from the cellar to circulate through channels in the stove body itself and emerge, toasty warm, out into the room. These days of renewed interest in wood heat are seeing a number of inventors applying for patents on similar devices—either freestanding fireplaces such as Franklin's or a number of heat-circulating fixtures for new or existing fireplaces.

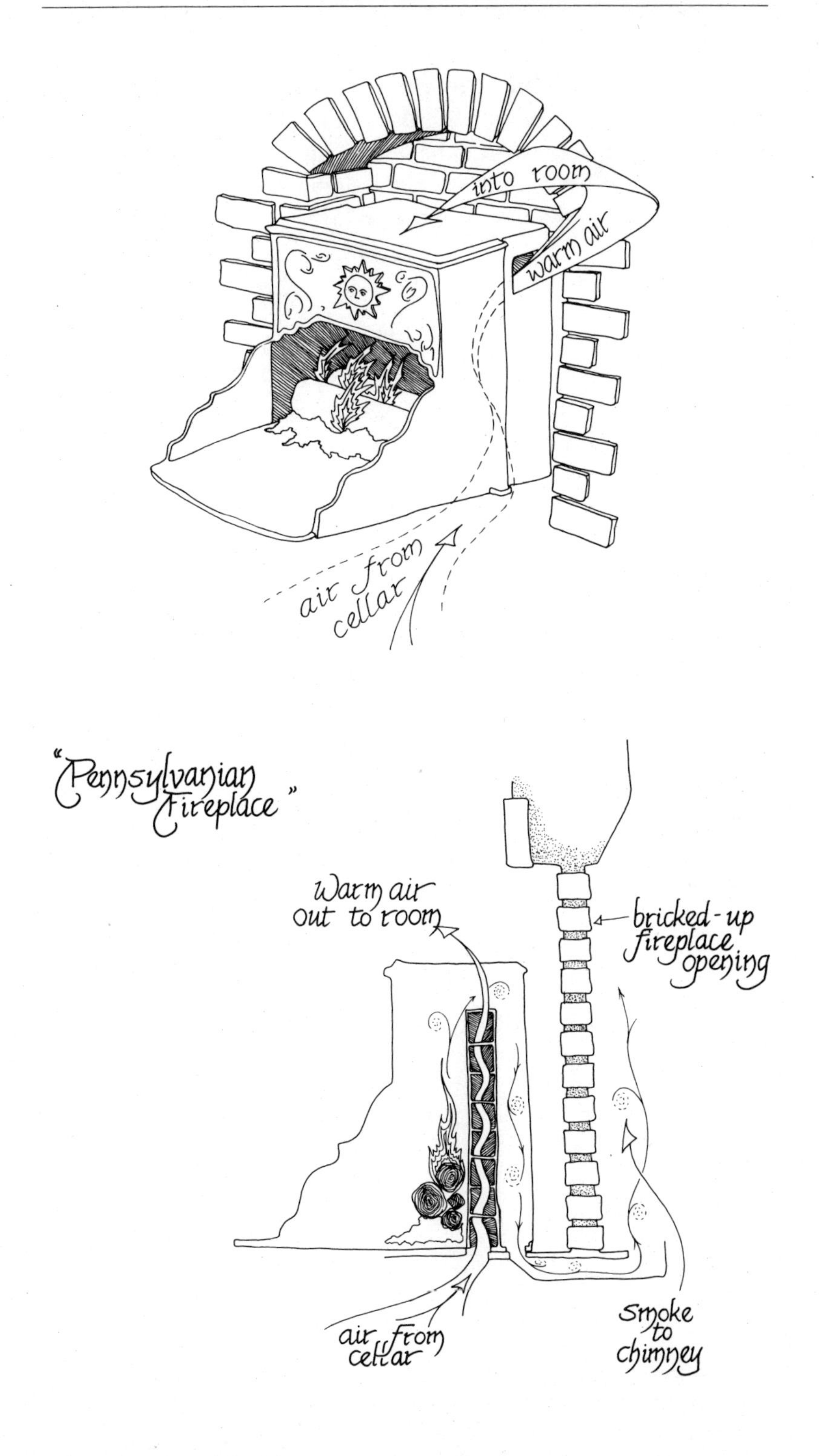
into room
warm air
air from cellar
"Pennsylvanian Fireplace"
Warm air out to room
bricked-up fireplace opening
air from cellar
Smoke to chimney

Few are much more than variations on "Poor Richard's" initial innovations, which Franklin never patented himself. He stated that they were for the benefit of all men, so should be free for all to use. Which is fair since some sources claim that Franklin's ideas weren't original at all, but "borrowed" from earlier inventors.

King Coal

Ben also applied his ingenuity to a result of the diminishing wood supply, a growing dependence on coal. At first it was sea coal, shipped over from Europe at considerable cost. But the high price of imported coal sent the resourceful Americans scrambling for local substitutes. Woodcutters went far afield after charcoal. Acre on acre of trees was cut down, the split logs piled in ricks, then covered and let smoulder. In the process (similar to coking off coal) the water vapor and volatile gasses are driven off. The result is nearly pure carbon, thus lighter and more compact. It was less costly to ship the increasing distances between the forests and population centers. But then the Appalachian coal deposits were discovered. Ben Franklin's next heating innovation—a generation after his wood stove—was a coal burner using the "downdraft" principle. He reasoned that if the flow of air through the fire could be reversed—oxygen-laden air coming in from the top—it would draw volatile gasses down through the hottest part of the fire, the coals, where they would

Acres of timber...
put into smoldering charcoal mounds.

burn completely. His downdraft stove was heir to sooting up, though it did try to solve the creosote problems we'll discuss when we get to the modern "airtight" stoves. However, burning updraft or down, coal quickly became king in the rapidly expanding America of the nineteenth century. Coal, of course, is nothing but wood and other vegetative matter that, over the eons, has been compressed to a nearly pure mineral form. Many coal seams bear the imprint of the leafy fronds of the tree ferns that lived during the Carboniferous period, the name literally meaning a time in which sunlight was accumulated, later to be compressed into carbon.

Coal contains more energy, more heat—more stored sunlight—per unit of weight than either wood or charcoal. As such it was far more economical to produce and transport, and with coal, more energy could be stored in a smaller space. Coal-burning locomotives and riverboats could go longer between fuel stops than the wood burners. Centers of population and industry could be located farther from the mines than from the earlier fuel sources, the forests.

When local mining reduced the price of coal, it replaced wood, and the urbanization and industrialization of America began in earnest. At the same time that this power source permitted the concentration of population in towns to serve industry, it permitted the agricultural portion to move onto the relatively treeless Great Plains. It was coal that powered the trains that brought settlers to the West in meaningful numbers. By the turn of this century it was cooking the meals and heating the houses in Kansas City and points west. Wood was used to build the homes and churches and commercial buildings of a growing America.

The Reign of Oil

Coal remained the dominant energy source for a relatively brief time. Though available in our country in great quantity, it was soon replaced by an even more concentrated form of sun energy, petroleum. Being liquid, oil is even easier to obtain and transport than coal. And it is comparatively easy to break down into even more highly concentrated energy forms such as fuel oil for your home furnace and gasoline for your car. And just as coal permitted easier transport and storage of energy than wood, petroleum products permitted still more flexibility in industry and life styles. Coal or wood could never power a jet bomber, a submarine or modern automobile. The sheer bulk and weight of these solid fuels would prohibit cities in temperate climates from growing to the size of a New York or Chicago. Without oil, we never would have reached today's and tomorrow's levels of human overpopulation. Both the chemical fertilizers and pesticides on which modern superproductive agriculture is based come from petroleum or its companion resource, natural gas. Which is to say that the discovery of oil has been a mixed blessing.

Further, during the heyday of cheap oil, we've let our other energy resources languish. The coal mines have been let go to ruin and until just the last few years the most visible use for wood was in veneer to panel dens and recreation rooms in which the fireplace was an ornament, the blaze mainly for entertainment.

Wood Heat and the Environment

Well, those days are gone, and unless the still questionable claims of the proponents of nuclear, solar and other high-technology energy sources come about, they are gone for good. Petroleum is in limited supply and is bound to run out eventually—hopefully not before the powers that be realize that it is too valuable to burn, and is better saved to make plastics, medicines and such than to go up in smoke. Coal is still with us in great amounts, but available only through deep or strip mining, the former costly in money and miners' well-being, the latter in money and environmental damage. And from whatever source, coal releases sulfur and other pollutants into the air. Both these fossil energy sources—these wonderfully concentrated forms of stored sun energy—will remain available for many years, though at ever increasing cost. Which brings us to the main point, the economics of wood as a fuel. As of this writing, in the depth of winter 1975/76, wood is easily the most economical energy source for perhaps one-third to a half of America's families, those living in rural, small town and suburban areas near forested areas. Let's examine the facts and figures.

First, how about air and other environmental pollution; wood fires smoke, and smoke pollutes, right? Well, yes, but it's as close to "good," natural pollution as there is. A wood fire is

nothing but a speeded-up version of natural decay, as previously noted. In the forest, every tree will eventually fall, and as it moulders away it will release water and CO_2 into the air, leaving its minerals in the soil; it's the same process as having a grate fire and putting the ashes in your garden, it just takes longer in nature.

You might suspect that intensive harvest of a woodlot would increase the amount of by-products of wood decomposition that are released, on the assumption that burning destroys faster than the natural process would. This too is a misconception. (Though the wholesale deforestation by fire of large areas, such as is being done in the interior of Brazil, may upset the natural balance. Not only are many square miles of oxygen-producing trees being taken out of "production" in Brazil's jungles, the fragile tropical soil is being exposed to the elements and the huge doses of carbon dioxide being released into the atmosphere by the fires may be affecting the climate. One theory holds that the net loss of oxygen and gain of carbon dioxide contributes to a global cooling trend, which in turn creates such problems for man as the droughts in Africa and possible return of dust bowl conditions in our own heartland.)

However, if we operate our own woodlots to have a continual sustained yield, and stay with native tree species and not try to import exotic foreign trees or breed superfast-growing hybrids (as some forestry people are doing in the Southern pulpwood plantations), a given amount of acreage will produce just so much vegetation per year. No more than the naturally received sun permits. You can "clear-cut," removing all the wood in one swoop, then wait 25 years or more for the next cutting. Or you can harvest a cord of wood per acre every year forever. Either way you are removing only the amount of wood—the stored energy—the land would turn over naturally in the given time period. If you or I don't harvest the wood using selective forestry techniques to produce the most good lumber and firewood of the desired tree species, Nature will harvest the same amount using her own methods of selection. Clear-cutting destroys the beauty of a forest and if improperly done can produce severe land erosion and other problems à la Brazil. But not even that unlovely technique can make the woodlands produce more wood than they would on their own hook. (Restoration of the minerals to the woodlands may be a problem over time, though trees "mine" minerals up from deep underground; more research is needed.) However, alone of our fuels, wood is the only one that is self-renewing, the only one that is nonpolluting. Put another way, burning wood that's harvested sensibly produces no more environmental effect than would the natural decay process.

That's assuming you and I and all the other wood burners apply the same sustained harvest principles to our own woodlots that the big lumber companies do. I have more than one quarrel with the plywood and paper makers, but one thing they do is keep that wood coming on. If we permit too much more of America to be deforested and turned into Los Angeles-type urban sprawl our woodlands could quickly become less than self-sustaining. It would take more than Ralph Nader and an environmental movement to save the forests if very many people were going cold in their homes. And if you want examples of what happens when the forest is overcut, the world is full of them. The final tally is yet to come on deforestation in Brazil. But China and India are treeless by and large, due to centuries of too many people living among too few trees. It's still debatable just how much of the world's deserts are products of climatic change, and how much is due to overcutting by man and overgrazing by his herd animals. But until the nation of Israel began the job of replanting them on their former hillsides—gone to desert—the cedars of Lebanon were only a part of biblical history. I wonder if there would be anyone left to replant the redwoods of California or maples of New England if they were cut out? There are plenty of woodlands now to warm us. Let's all stay politically alert, and keep America green — as well as warm.

The Economics of Wood Heat

If we do lose our woodlands, it will result from someone's selfish and shortsighted effort to make a quick profit. So, we'll examine the dollars and cents (sense?) of wood heat. First, the larger picture of capital investment and jobs for people from our several energy sources: Each gallon of fuel oil or ton of coal you burn is a result of an investment of millions of dollars in refineries and mines, crude oil tankers and coal trains. Practically no labor is required in oil production after the well comes in, and the distribution from wellhead to refinery to your home tank produces relatively few jobs. The main economic result of oil-domination of our energy system has been to make very few men and institutions very rich. Today most of them are in the Middle East, though a few Texans are still doing OK, I hear.

Coal isn't much better. It produces more jobs, but the hard work and dangers of a miner's life plus the dust, grit and grime of the entire delivery process don't make the sort of jobs that encite much envy. Unlike oil, the wealth of much deep-mined coal stays closer to home. But the mine operator's home still occupies the top of the hill, the miners live down in the hollows. The economics of strip mining are more similar to oil production. Huge machines roar in, remove the soil, then gobble the coal seams and move on. Perhaps a few local heavy equipment operators have good jobs for a while. But most of the money goes to the huge company that owns the equipment and mineral leases—more rich folks getting richer.

Now, I'm not necessarily going on like this to criticize the oil or coal companies. Though, like yourself, I'd bet, I do enjoy reading about the criticism they are getting from other quarters. And I'm not grumbling about America's highly successful economic system or the folks who accumulate the wealth it produces. Louise and I had our shot at that life, we didn't think the rewards were worth the effort and chucked it in large part.

But that's personal preference, not a condemnation of anybody or anything. I'd suggest that switching from oil or whatever to wood heat actually supports a healthy capitalism and wealth accumulation. Not by some distant millionaire, but by the hardworking local guy who needs the money badly enough to go out into the woods and work up firewood. Of course, ideally that person is yourself; property taxes aside, the "price" of wood harvested from our own woodlot costs Louise and me only the price of a little gasoline for the machinery plus a lot of sweat. As Henry Thoreau said, "Wood heats you twice, once when you cut it and again when you burn it."

But even if you purchase wood, most of the cash goes directly to a hardworking crew as pay for time spent. Getting in firewood is almost entirely labor. The basic capital required is for a truck, a saw and a splitting maul. And woodcutting is a one- or two-man enterprise. Exxon isn't interested. It just seems to me that in these parlous times your own hard-earned dollars are better spent paying for the honest work of a neighbor than heading off to Chicago or Dallas or Addis Abbaba or wherever to fatten the bank accounts of people who are already driving Rolls Royces. Wood heat as global politics? Well, it's worth thinking about.

Personal Economics

So much for the big picture. How about the smaller one featuring your pocketbook and mine? Quite simply, in terms of strictly cash outlay, wood is the best energy bargain available if you live reasonably near the woods so that transportation costs are minimal. There is a great deal of fetching and hauling to do, however, and if you put a great premium on your time, or if your job is such that you want home time to be mainly relaxation, wood heat may not be for you. Let me try to put some rough figures on it, based entirely on our own preferences, experience and prejudices.

First, let's try to make a comparison of costs versus heating value for the major energy sources. I've read several books and articles that attempt to equate costs of wood vs. oil vs. electric heat down to the penny. This is silly because of the tremendous variation in costs of each fuel in different sections of the nation—differences in heat values among different grades of each kind of fuel and in the efficiency of heating plants. For a rough set of numbers applicable to our part of New England the winter of 75/76, see the chart "Comparison of Heating Fuels." We've tried to figure the amount and costs of other fuels that would put out the same useful heat as a cord of well-dried beechwood. A cord of beech, weighing in at about two tons, is roughly equivalent to a ton of hard coal, 200 gallons of #2 fuel oil or 4,200 kilowatt hours of electricity.

At current prices here in our country town, the wood would cost $40, the electric heat about $60, the coal $75 and the oil $80. In the big city nearest us, Boston, a cord of wood, delivered, costs at least as much as its oil equivalent, but much more than electricity, even here where electric power is highest in the nation. In the hydroelectric areas of the Pacific Northwest, electricity is a better bargain still. But in the anthracite coal regions of Pennsylvania you can get fuel that is almost pure carbon for less than the price we'd have to pay for a cord of wood. And I

COMPARISON OF HEATING FUELS

A Fuel	B Amount	C Weight	D Gross Btu Output in millions	E Efficiency of appropriate heating unit	F Net Btu's in millions (ExF)	G Cost	H Equivalent Costs (BxG)
Dry Beech Wood	1 cord	4,000 lbs.	28	50%	14	$40/cord	$40
Anthracite Coal	1 ton	2,000 lbs.	23	60%	14	$75/ton	$75
No. 2 Fuel Oil	200 gal.	1,500 lbs.	22	65%	14	40¢/gal.	$80
Electric Service	4,200 kwh.		14	100%	14	1.5¢/kw.	$62

NOTES:

D: Btu = British thermal unit, the amount of heat needed to raise one pound of water one degree fahrenheit.
E: Efficiency = amount of heat generated retained in heating system (rather than going up the chimney) by a good wood stove, oil burner, modern coal furnace and the 100 percent efficient electric resistance-heating units. Price of the latter lowest rate for large residential consumers.
G: Figures based on prices in Central New England the winter of 1975/76.

#2
#2
#2
#2
200 Gallons Oil
One Cord Wood

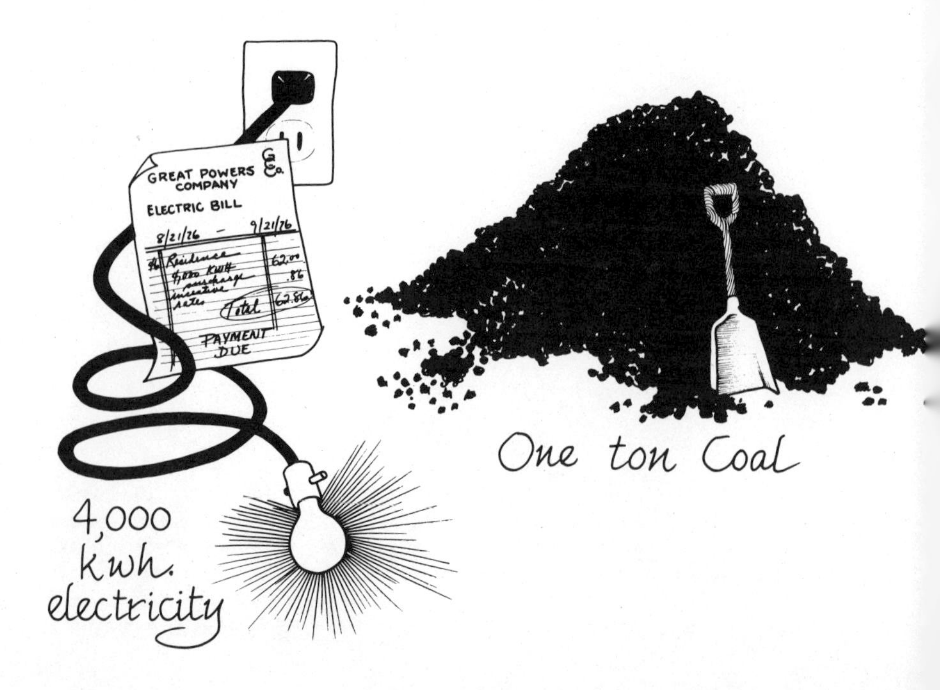
GREAT POWERS COMPANY
ELECTRIC BILL
8/21/76 – 9/21/76
PAYMENT DUE
One ton Coal
4,000 kwh. electricity

could go on. The point is, actual price differences will vary around the nation. You'll have to figure for yourself. But in general the rule holds true that a cord of hardwood, a ton of hard coal, 200 gallons of #2 oil and 4,000 plus kwh. of electricity will turn out roughly equivalent amounts of heat.

What about the economics of bringing in your own wood? The standard cord is a stack of four-foot long sticks piled up in a box shape eight-feet long and four-feet high. As mentioned above, weight of a good-burning hardwood such as hickory will exceed two tons per cord. A soft pine will run a bit over one ton per cord. Of course, to get a cord you must cut down a tree, saw it up, then split the larger pieces. Back when they had nothing but ax, saw, and splitting wedge, they figured that a good man could put up two cords per day, from tree to finished pile. Those were hardier days. The best I've seen in this modern age was two men, working at a steady pace with the much faster chain saw and a four-wheel-drive truck, getting five cords cut during a good day. The larger sticks still had to be sawed to stove length, 12 inches, which adds in time to put three more cuts along the entire face of the cord. Plus, splitting and hauling in were left till later, a good way to get a bit of exercise each evening during a sedentary winter.

Let's see now, those five cords would cost $200 if purchased today. Figure, what with breaks and a long lunch hour spent hunting pheasant in the cornfield, our woodcutters got in an honest five hours of work per man. That would make the "wage" come out to $20 an hour. Not bad. Even if we admit that another day's work would go into cleaning up and chipping the slash as well as stacking and splitting over the winter, the pay still comes out to $10 an hour. Of course, we could value their time in terms of cash *not spent* for the alternative fuel, heating oil. That would bring us back up to the $20 figure. Then, we can add on the money *not spent* in an alternative form of physical exercise or recreation. What does it cost per playing hour to belong to a golf club, $5? That's a wage of a quarter of a hundred dollars each 60 minutes. If we cut wood instead of renting a light plane, the imputed "wage" could jump to $50 an hour and more. What recreational exercise do you pay to enjoy?

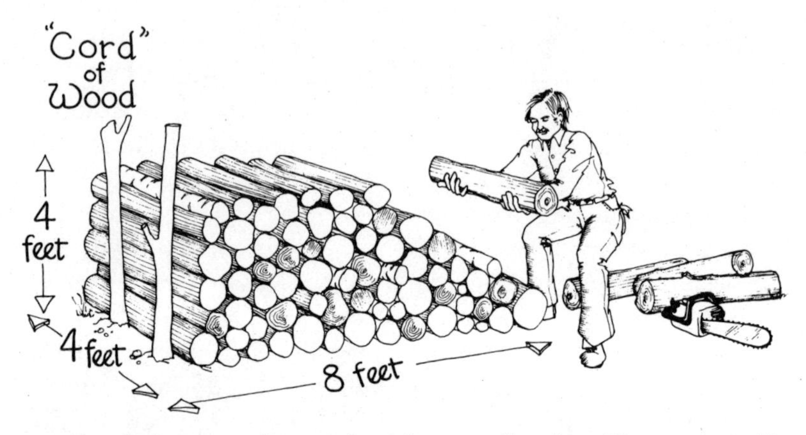

But let's never forget that the woodlands will remain self-sustaining only so long as we don't exact too much of a "wage," imputed or otherwise from them. I'll say it once more—let us wood burners insist the loudest that we all keep America green.

CHAPTER TWO

Chimneys and Flues, Old and New

The least interesting part of your wood fire is the chimney. However, it is probably the single most neglected or ignored. So, let's start with that all-important tube that keeps the fire burning and your room clear of smoke. When a wood fire burns it generates heat and water vapor and several gasses, mainly carbon dioxide. If the fire is young or the wood green, combustion will be incomplete and unburned carbon and assorted oils and resins will come off as smoke. Of course the fire consumes oxygen as it burns, so a fire just let burn in a closed space can make it uncomfortable, dangerous and even fatal for us oxygen-breathing creatures. Shortly after he discovered fire, early man must have found ways to get the smoke out and fresh air into his cave or animal-skin tent or whatever. Hot air rose then as now, so ways were devised to guide it up and out of the dwelling.

Probably the most familiar example of a so-called "primitive" smoke pipe is the tipi of the American plains Indians. These nomadic tribes followed the game in their seasonal migrations, moving the cone-shaped shelters of buffalo or deerhide and erecting them on frameworks of fresh saplings at each camp site. The structure had a closable entrance flap at ground level and one or more adjustable flaps at the top to regulate heat and smoke flow. In cool weather all flaps would be opened and a small fire built in a stone-lined pit, normally in the center of the tipi. Like most fires it would be smoky at first, but as it burned down and generated a bed of coals to assure quick and complete combustion of the wood, smoking diminished and the flaps were closed till only enough air came in to keep the fire's gasses moving upward so that smoke and fumes accumulated in the high peak of the cone.

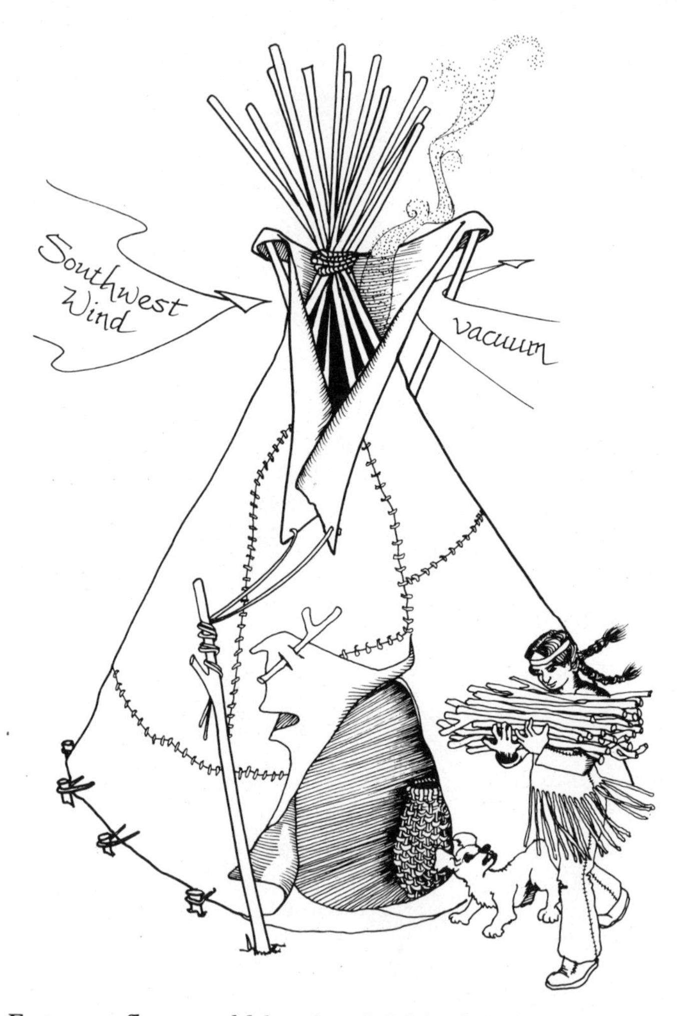

Entrance flaps could be closed fairly tight and the top flaps were arranged as a sort of chimney extension—aiming downwind and blocking the smoke hole itself from the direct path of the wind. In this way a smoke eddy and a slight vacuum were created at the flap—which acted very much as a bird or airplane's wing does in generating the vacuum that lifts its weight into the air. This effect kept the smoke and hot gasses flowing more or less continually out the smoke hole. At the same time, fresh air was being drawn in at the hole's edge to replace what was escaping at the center and along the flap. This cooler air immediately fell, but by the time it reached the occupants it had warmed from contact with the tipi sides and the column of hot air rising

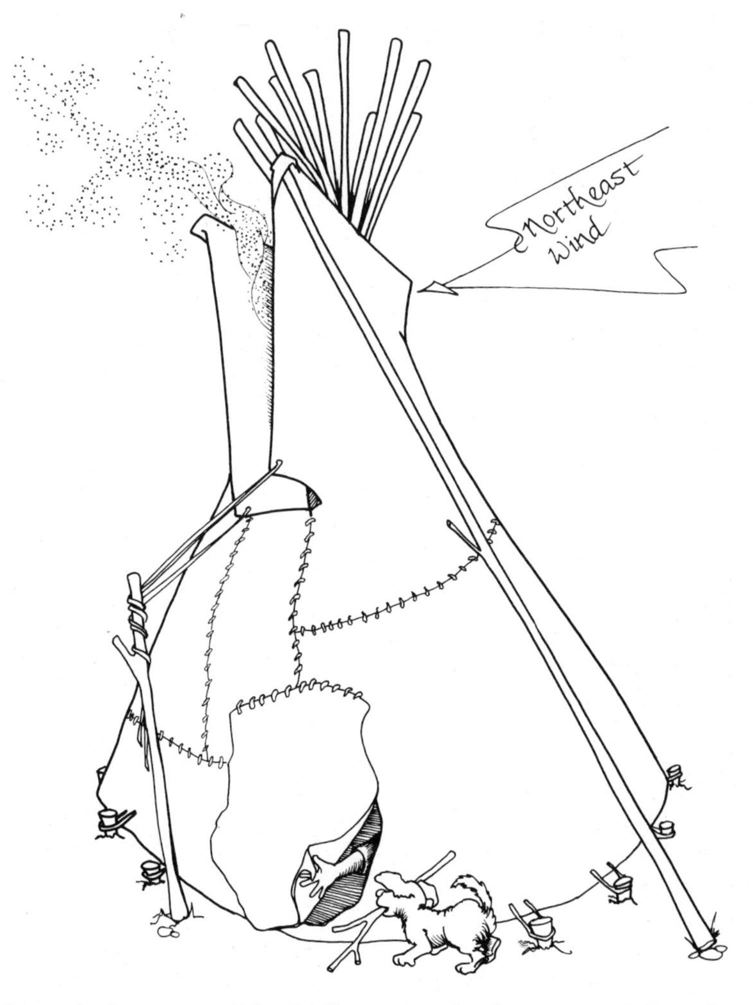

through the center. The Indians were further warmed from heat absorbed by the earth floor of the shelter that continually radiated upward.

The relative efficiency of a tipi in evacuating smoke from the dwelling place may have been a fortunate accident that happened when the cone-shaped design was developed through intuitive knowledge of aerodynamics and thermodynamics. The more typical shelter of preliterate peoples in chilly climates is a dome-shaped structure of animal-skin over a pole framework or a similarly shaped hut of whatever building material at hand — mud and brush or stones; the Eskimos used ice. But by all reports, the smoke from the central fire was evacuated

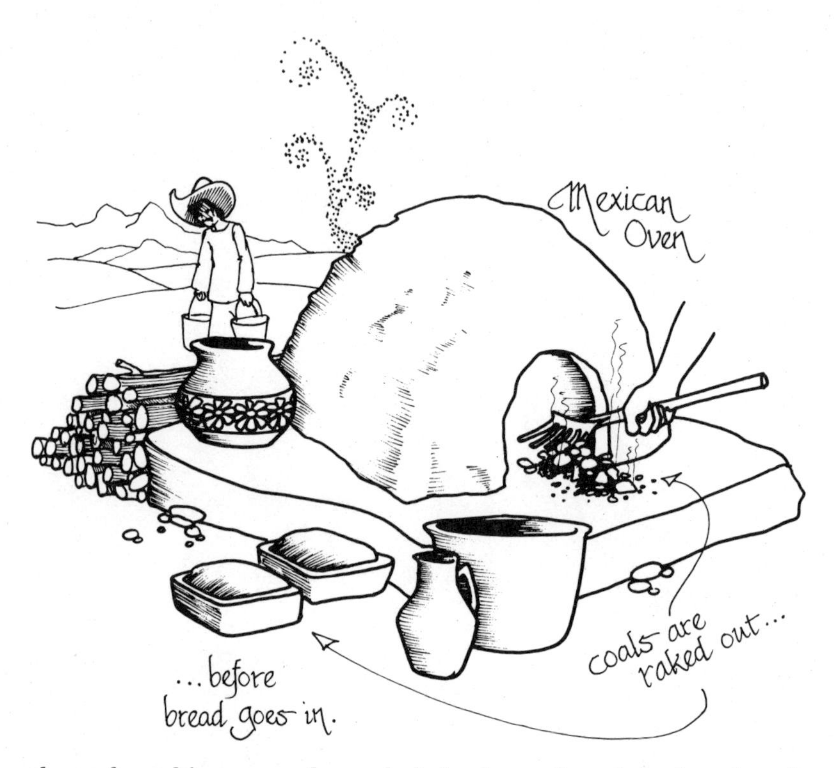

through nothing more than a hole in the roof, and the interior of the shelter was pretty smoky.

Some early peoples came up with wood-heated devices that wouldn't smoke you out of the place, but all the ones I know of were for cooking. The Mexican oven, still in use in one form or another throughout much of the less developed world is little more than a dome-shaped structure of adobe or mud and rock with a rock or brick door. Some types have a smoke hole in the back. They are heated by being filled with hot coals or having a fire built in the mouth. When the oven is ready, the coals are raked out and the bread is put in on the floor where it is baked by heat already absorbed by the oven walls and floor. A somewhat more flexible cooker is the Chinese oven, presumably an early development from China. The idea became mildly popular in the United States back in the 1940s when backyard barbequing was first becoming widespread and you occasionally see one in the backyard of a suburban house. (The oven was often used as an incinerator before air pollution became a severe problem.)

The Chinese oven is nothing but a low chimney with a door

Chinese Oven

at the bottom and a steel lid that has a hook welded on its lower side at the top. Size varies, but waist high is most common. You build a wood fire in the bottom of the device, regulating draft by opening or closing the bottom door (or by regulating a draft control if one is built into the flue opening covered by the steel lid. Once the walls of the chimney are hot, the fire is let go out or die to a bed of hot coals, depending on what is to be cooked. A haunch of meat, several fowl or a pot of stew is hung from the hook on the lid bottom, the lid goes on, and the meal is cooked by a combination of radiated heat à la Mexican oven and upward radiating heat and smoke from the coals. It is a full-time job, to keep heat constant when cooking with this device, but the result has the flavor and juices sealed in by the surrounding heat and the coals give the food a wonderful smoky tang. (We've never built a Chinese oven from scratch but have gotten much the same effect by hanging a piece of meat in the preheated flue of an outdoor fireplace. It is a great way to cook a leg of lamb or goat or a ham while finishing off the maple syrup—boiled all day over the outdoor fireplace but cooked the last crucial hour or so on the range inside. Potatoes can be baked just by putting them in the bed of ashes—covering with foil if you're finicky about a bit of wood ash on the plate.)

The Development of the Flue

As the human society changed from nomadic or semi-nomadic hunting or foraging orientation to an agricultural one, life became more settled. Villages became towns, then cities. Better and better ways were developed to guide the smoke from heating fires out of living space. First, homes became more substantial, built of wood or stone, thus less drafty and more capable of retaining heat. It was probably in the relatively balmy climes of southern Europe—Italy by all reports—that the fire was first moved from room center to a far stone wall. A smoke hole was put in the wall just above the flame. The wood heated not only

First, a hole ... Then, a hood...

the room air by radiation, heat was absorbed by the walls themselves, and continued to emit warmth long after the flames were out.

Next step was to construct a hood on the wall to gather smoke, and then a pair of sides were added, extending from a hood's edges down to near floor level—all to contain the smoke and guide it up to the smoke hole. The next idea was to further contain smoke by recessing the fire into the wall itself. Thus was born the fireplace. But somewhere along the line came the major breakthrough—the flue or chimney, the latter term originally

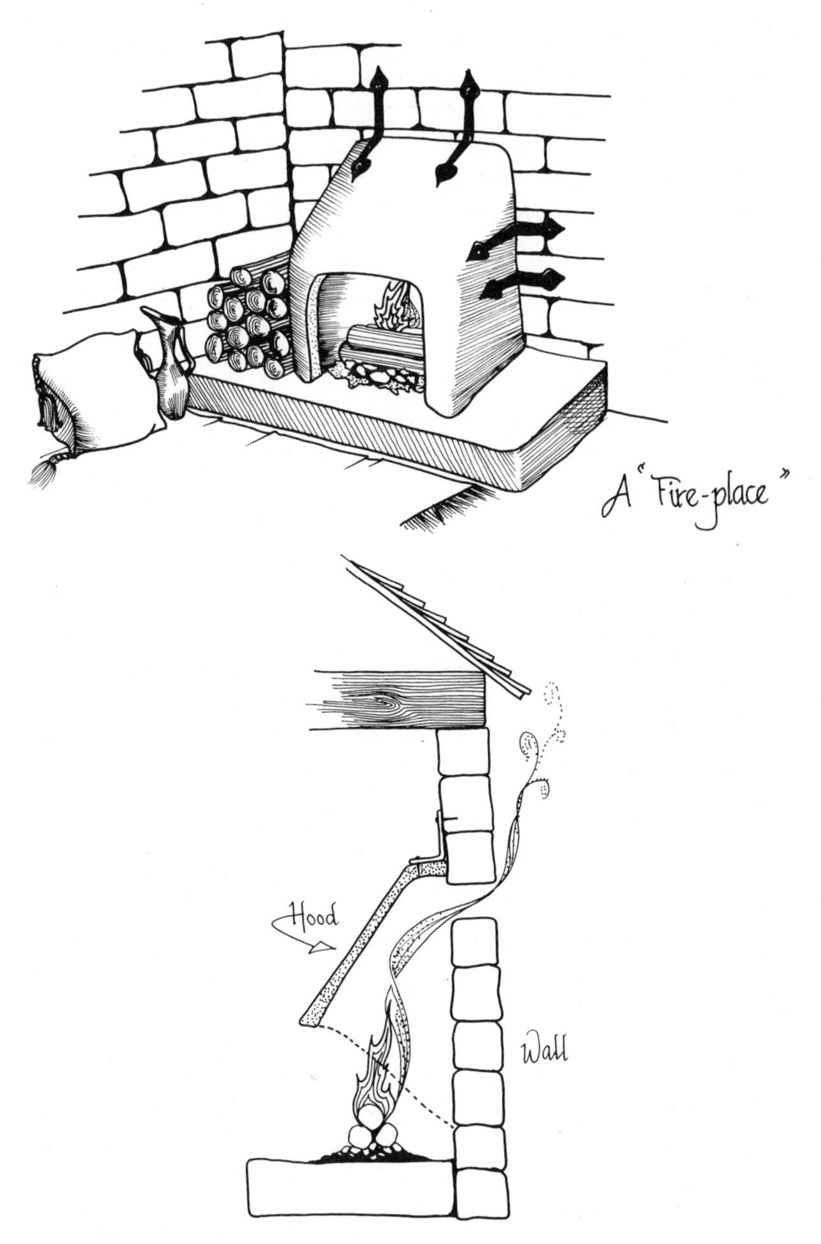

referring to the entire apparatus from fire base on the floor to the top of the smoke outlet. The concept was a feature of the homes of the affluent in Britain by the twelfth century.

With a flue or not, hot air rises, and smoke from the heat-generating fire is carried along with it. But the vapor must stay hot to keep going and the hotter it remains, the faster it will rise.

In the open center-of-room fire, smoke cooled greatly before it got to the smoke hole, and the development of hoods and such served mainly to bottle up most of the smoke till it found its way out. Whoever discovered the flue realized that if you keep the smoke hot and at the same time guide it (of necessity up and out) away from the living spaces, you will breathe clean air despite the fire.

And that is all a flue is, an air guide that takes advantage of the tendency of hot air to rise by keeping it warm and guiding it wherever it is wanted. The longer the air is kept warm, the faster it will rise; thus, the higher the flue, the better will be the exhaust, the stronger the draft. If you've ever wondered why a fireplace fire takes a long time to get going, and why it smokes at first, it's because your flue is cold and is chilling the smoke. Once warmed, the flue no longer cools the rising air, thus the fire "draws" well.

For best draft, the flue should offer a minimum of impediment to the upward-flowing warm air. So the chimney or stack should be as straight-up as possible, as high as practicable and with the top free from interference due to eddies or other barriers to smooth air flow caused by nearby buildings, trees or your own roof. The interior should be smooth as possible and airtight. Cracks can let cold air in and sparks out; the former can cause poor draft, the latter can burn your house down. So let's get right into inspecting and maintaining existing chimney and flues. Then we'll build a couple of types you might use temporarily while a permanent chimney is being built at home or perhaps build for intermittent use at your hunting camp.

Inspecting and Maintaining Chimneys

To last and work safely, chimneys must be of materials that are impervious to fire or weather damage. Since all or most of a chimney may be hidden from view within the structure it heats, it must also be designed and built to hold up safely for many years with a minimum of maintenance. So, most chimneys are made of vitrified ceramics such as brick, of formed concrete lined with ceramic or of stone. Multi-walled, insulated steel flues that are prefabricated have come into wide use the last few years. All make serviceable, long-lasting flues. Let's begin with permanent flues, specifically, with the old flue you may be reactivating after

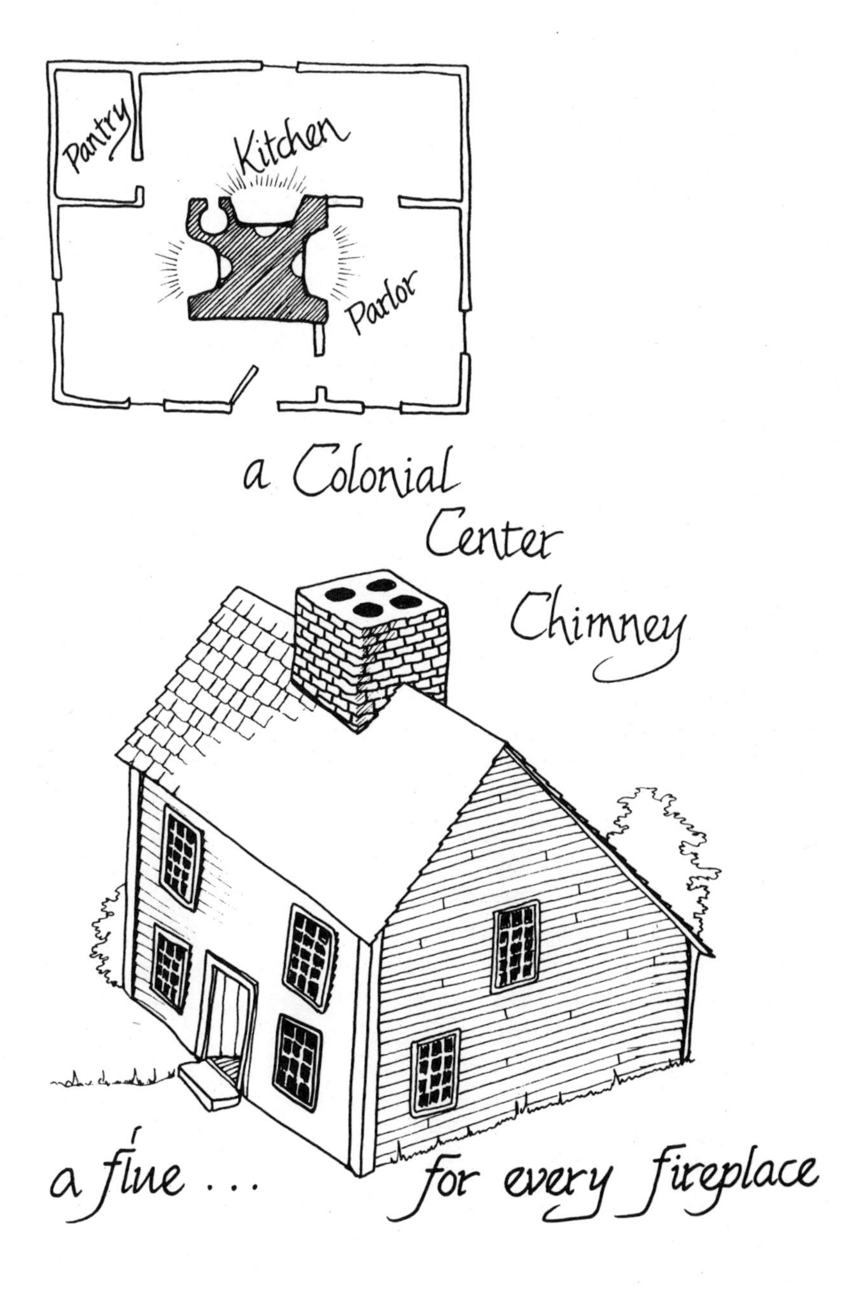
Pantry
Kitchen
Parlor
a Colonial Center Chimney
a flue . . . for every fireplace

years of disuse, such as the brick and stone chimneys that Louise and I found on the farms we've lived on in the past. If they've been protected from water damage many old flues are basically sound; ours have been and I hope any you may be putting back into service are, too. Here's how to tell, and how to fix any problems.

Locating the Flue

Start up on the roof; that is where most chimney damage takes place and where nearly all of it originates. If your house is an old center-chimney colonial you may find as many as five flues in the single chimney. The monolithic brick structure heated up and radiated heat through the entire house during the winter. Also the early Colonial governments levied a chimney tax, so the crafty Yankees put all their flues into a single chimney. The number of openings in the chimney top will tell you how many fireplaces there are or were below. In our last farmhouse the old fireplaces had been torn out and a simple buttress put under the flue when the more "modern" and efficient cast-iron stove became popular. Many houses we know have fireplaces bricked up, with only a stovepipe running into the flue; still other fireplaces have been entirely sealed away behind the walls.

South of New England the flues will often be hidden in the walls of brick or stone houses, sometimes with the chimney top removed and roofed over. Here you'll have to pound around with a mallet and listen for the hollow sound that betrays a flue. If your older house has had its flues sealed to be put out of commission permanently, you may be better off building new ones. I'd call a good mason if repairs or rebuilding flues in an old house entailed much more than a bit of patching or the makeshift devices we'll get into later. Neither Louise or I are experts on early house construction, and unless you are, I'd advise calling in someone who knows the subject cold. Don't put in something that could be dangerous or damage the appearance and value of your house.

As early as Ben Franklin's time, houses were being built specifically for stove heat and central steam heat was available for homes well before 1900—indeed, had been in use in public buildings as early as the mid-1700s. Here, the flue or flues can be located anywhere in the house—back, sides or anywhere inside. The place we live in now was built in the late-1800s and is fairly

typical. There is a single central flue and into it exhausted a big coal-fired hot water boiler plus three wood stoves: the kitchen range and bedroom stoves on the second and third floors. The coal furnace was replaced with an oil burner some time back. We hope to replace that with a multiple-fuel wood/oil burner one of these days (the oil as a self-tending backup for when we are away in the winter).

Flue Capacities

If your flue is used with any sort of central heating device, most experts will advise you to build a separate flue for wood heat. It's true that a fireplace and modern furnace should not operate on the same flue. The great amount of draft a fireplace requires can interfere with the furnace's operation. But an efficient stove or stoves will not—particularly if they put out enough heat that the central unit does not operate at the same time they are on. More on that later.

If you find flues with no apparent fireplaces and no connection to a central furnace, you likely have a chimney that was

old flue
ready
to re-use
...or it
may
be hidden under
the wallpaper...

built for wood or coal stoves. You may find sheet metal covers over the holes in the walls—many had decorative scenes printed on them. Plates may be buried under wallpaper, or the holes may have been plugged, plastered or cemented shut. Get out the mallet and pound walls. The openings will usually be somewhere near chairback height or higher.

Evaluating the Masonry

Now that we know what you have in an old flue, let's get back up on the roof and finish the inspection. There should be a good cap around the flue, a solid mortar coating that prevents water from getting into the top ceramic joints and pushing out the mortar. Use a screwdriver to probe around in the mortar joints up and down the chimney. If you can dig out small chunks or even a brick or two, it's alright. They can be replaced. If all the joints are rotten, and particularly if the bricks are crumbly, you'd best plan to rebuild at least the outer layer of the chimney top. Look inside; if the inside shows rotten brickwork you'd best plan to have the chimney rebuilt from top to at least a foot below roof level. If a flashlight shows loose and crumbly joints and broken or protruding bricks well down into the flue, I'd call a mason and be prepared to totally rebuild. The last central chimney (three fireplaces included) rebuilding job I witnessed cost

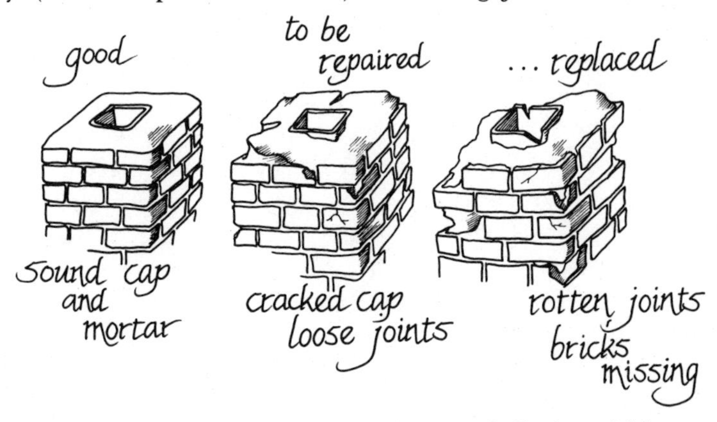

about $3,000 and that was a good many inflation-ridden years ago.

Assuming you can go ahead and use the chimney, check where it mates with the roof. This is a most critical part of the structure, where rainwater is kept out of the flue and the attic. There should be flashing; sheets of lead in older houses, copper in younger ones. One edge of the strip should be embedded in mortar joints or mortared directly to the masonry. Then it should run down to the roof sheathing, and out for several inches under the shingles or other roof covering. The whole joint may be covered with a thick coating of tar. If the flashing is missing or in bad repair, get ready to fix it.

Now, repeat the inspection of joints and brick, stone or whatever on down to the foundation. Get into the attic, cellar and any other locations of access. Look for loose mortar as on the roof. If you see dark streaks that look as though sooty water has flowed out through the bricks, you're right. The problem may be lack of a sound cap; the bricks literally soak up water if you let them. Or it may be rotten joints. Your screwdriver will tell you which.

If the flue is located in a wall or outside a wall its foundation will be as good as your house's. Old New England homes with central chimneys have a large stone foundation built up in the cellar, and these seldom have more than a stone or two out of place. Most modern flues rest on a sound concrete or crushed rock foundation. But I've seen some middle-aged houses where fireplaces and flues were built on huge planks strung between a pair of stone buttresses in the cellar. Anything like that deserves some more support—a solid foundation or a couple of heavy-duty house jacks put permanently in place.

Now, unless the flue is obviously beyond saving, build a smoky fire in the fireplace stove, or light an oily rag and put it in through a stove hole. Plug the rooftop opening with an old pillow, a feed bag full of leaves or the like. If there are holes in the flue, the escaping wisps of smoke will tell you where. Mark them for repair. If the flue looks as though it has sprung a thousand leaks, I'd abandon it and build a new one, have it rebuilt, or perhaps keep it for ornamental value and give it a new life and lining as described later on. Remember, for each leak you can see, there are probably several hidden behind walls.

If your house is relatively modern and the chimney done by a good mason, there should be no severe problems. If there's a smooth liner in the flue, made up of round or square cylinders molded of fireclay placed on top of the other, you've a good, modern chimney. Most likely there will be two or perhaps more flues in the chimney, one for the central-heating furnace and one

Stuffing flue-top with straw-filled sack...
...to find leaks

for each fireplace or stove (s). About the only problem that crops up in the older versions of these chimneys is leakage from one flue to another. Often the joints between tile liners weren't mortared and in time interflue leaks developed. There is some slight chance that sparks could travel from a warm flue into a cold one, drop and start a fire. And the draft of both flues can be affected by the leaks. They can be remedied easily.

Repairing Faulty Flues

For fixing the outside of a chimney you'll probably need a small mason's trowel, a pointing trowel (like a bent screwdriver) for repairing joints, a bag or two of premixed mortar mix (not sand mix or concrete mix), and a bucket. First, dig out all loose mortar and broken bricks or other building material. Then mix up as much mortar as you think you can apply in a half hour. Wet down the chimney area you will be working on so the cement will mate well. Then use the pointing trowel to fill in joints. Pack

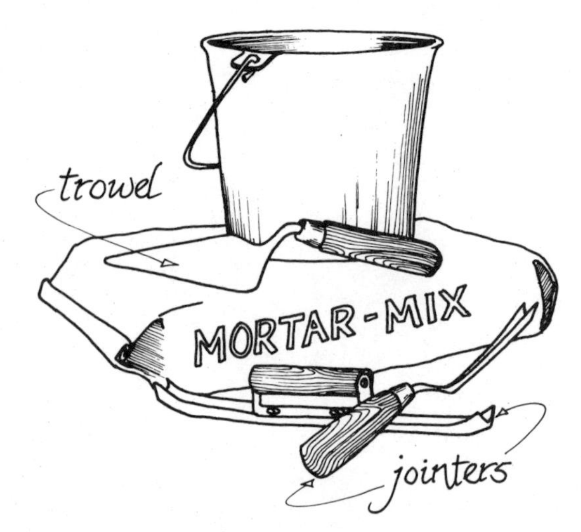

mortar in well, but leave a groove in the finished surface of each—protruding cement invites water in. Loose bricks can be mortared back in place unless they have gone soft.

The Exterior

The portion of chimney exposed to weather may be badly deteriorated, while the inside and part of the outside under cover are in good shape. Here, you can just plaster the entire outside. Wrap tightly with fine-mesh poultry netting, or use metal wall

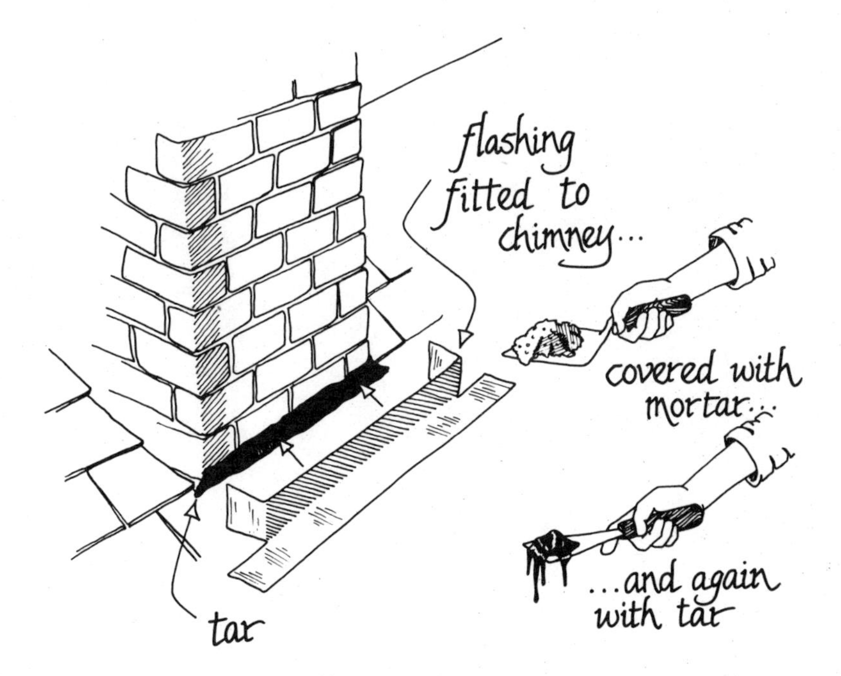

lathing. Just make sure it is wrapped tightly and as close to the brickwork as you can get it. Then plaster with the mortar mix. Do it in two coats, the first to fill in spaces between wire. The second, before the first has dried completely (12 hours or so later), to cover the wire.

If plastering a chimney, you will want to reflash between first and second layers. Get some copper flashing from a builder's supply outlet: a roll of six-inch-wide flashing should do. You'll also want a can of roofer's cement or tar and a putty knife. Where the chimney meets the roof, parallel with the eaves, ridgeline and run of shingles (horizontally) you will not want to pry up roof covering. Put a thick layer of tar along the roofing at the joint. Then cut a strip of flashing and bend as shown.

Slip bent strips of flashing in under the shingles, starting at the bottom. Press the flashing into the mortar. Now apply the second coat of mortar, covering the flashing on the chimney completely. Once the mortar has dried put a coating of tar on the horizontal flashing, covering it thickly and completely. If you've not been able to get strips in under shingles on the sloping joints, you should tar them also.

Reflashing a chimney that is not to be completely plastered is about the same job, but you tar along the entire joint in place of

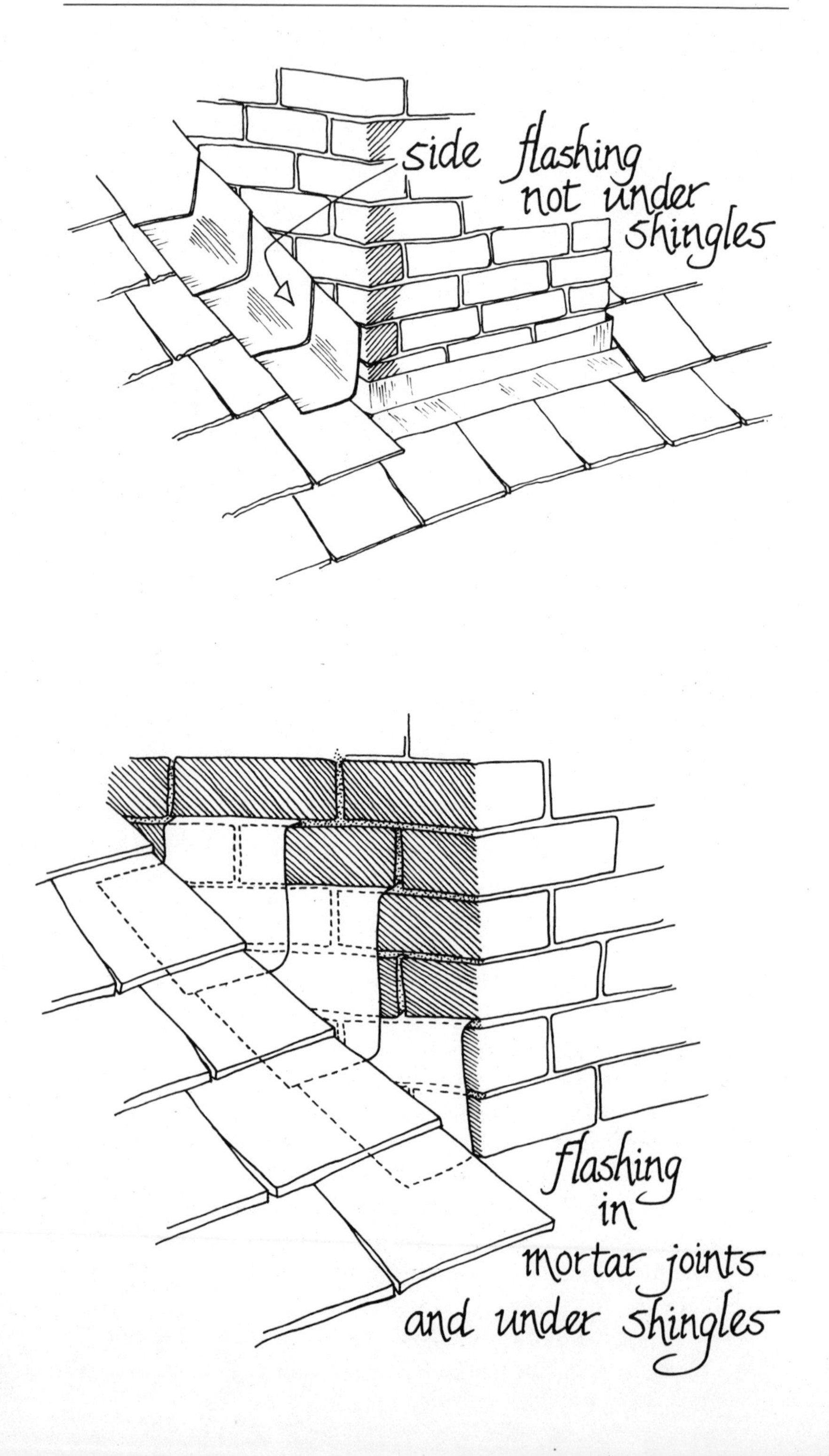
side flashing not under shingles
flashing in mortar joints and under shingles

concrete—a tar coat, the flashing, then more tar. If the mortar joints are loose enough you may be able to slip an edge of flashing into the brickwork and cement it in. This makes the best job of all.

All chimneys should have a cap in the form of a low cone with the top cut off to shed water. You can also put on a rain cover, an especially good idea if the inside of the flue is in bad condition. If you live in a particularly windy area, I'd suggest putting no more than a thin, flat cement cap on at first, running the flue for a while to see how the prevailing winds affect your draft. You may want to incorporate a windbrake into your rain cover. There are also a number of ceramic chimney pot designs that you can just mortar on. They help alleviate wind problems and also increase flue height, often a quick and easy cure for a smoking fireplace or poor stove draft.

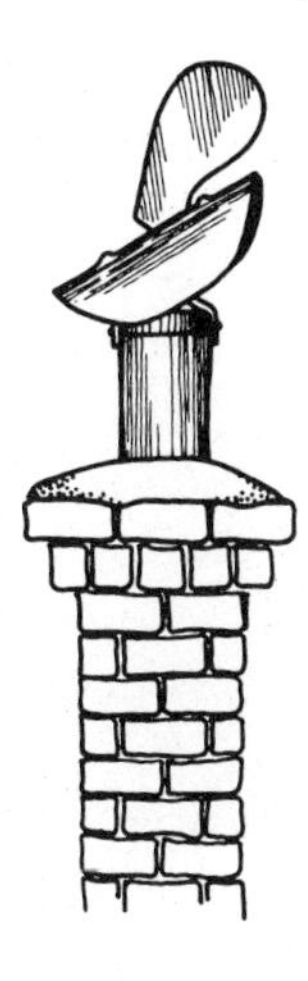

other, familiar
chimney-toppers

If your chimney is located on an outer wall so one entire side is exposed and in bad shape outside, you can plaster it up just like the top. The worst problem is fastening the chicken wire to the brickwork. I've found that if the brick and mortar is punky enough, sturdy electrician's staples can be driven in between bricks securely enough to fasten yard-wide strips of wire. Fence staples are too soft, as are the staples that come with staple guns, which are also too short. It takes a lot of climbing up scaffolding and much hammering, but a complete plastering job done in your spare time is a lot cheaper than a complete chimney-rebuilding job.

While we're up on the roof, better put some screening over the flue top. It will keep large sparks in and such bothersome chimney dwellers as bats and nest-building chimney swifts out.

The Interior

Reconditioning the inside of flues is chancy. Old chimneys that were used for years of wood burning will likely have a many-inch-thick coating of carbonized smoke deposits as an inner lining. So long as you keep it from catching fire, it is about the best lining you could wish for. But after years of disuse and exposure to rain, this creosote can come loose. Or you may have bad joints and rotten brick inside as well as out. It is a good idea to scour the inside of any old flue. Get a set of tire chains and enough rope to reach from roof to flue bottom. Hang the chains down the flue and pull them up and down repeatedly from top to bottom and all around. Anything loose will come off in time. You may have a few hundred pounds of junk in the bottom of the flue, but you'll know where you stand. Inspection of the inside is done most easily at night. Tie a strong lantern to a cord and lower down the flue. If the light reveals that any big chunks of masonry are missing, plan to abandon the chimney or line it with pipe, which I'll get to in a bit.

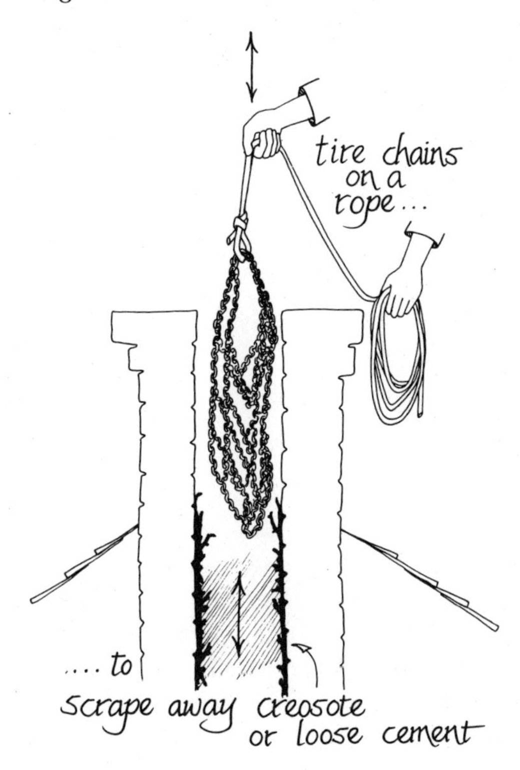

Recementing the Flue

If all looks sound, go down and light your fire. However, if there are loose joints, broken bricks (or uncemented joints between sections of liner in a modern flue), you have a messy job ahead. First get a bag of fireclay cement or furnace cement from a heating contractor. This is the same stuff you use to mortar a masonry firebox for a fireplace; heat hardens it up to a rock-like state. Next make a flue traveller. Best is a board an inch smaller all around than the narrowest section of flue. Attach ropes to each corner, as shown, so the plate will pull straight up the flue without tipping. Then enclose the plate in a bag of tough cloth filled with rags, straw or other loose material, again per the drawing.

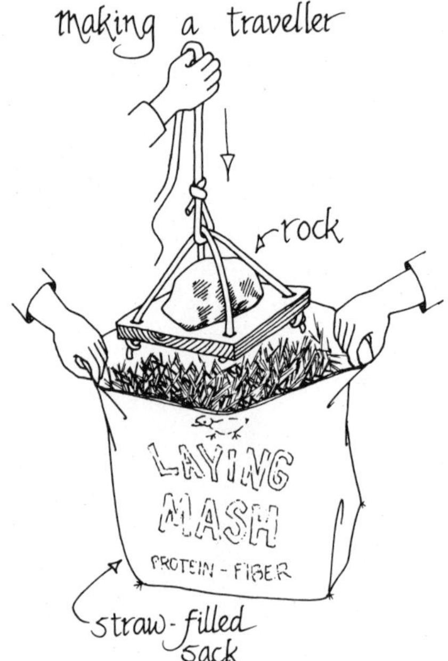

Now mix up the cement to a consistency of heavy cream. Attach a line to both top and bottom of the traveller, the top line running to rooftop, bottom one hanging down the flue, just in case the traveller sticks and has to be pulled down. What you want is to pull the traveller up to each joint and pour enough cement to fill the cracks down the flue. It's a messy job. You'll end up with half the cement at flue bottom, and the traveller will likely stick a half-dozen times before you are done. I've done

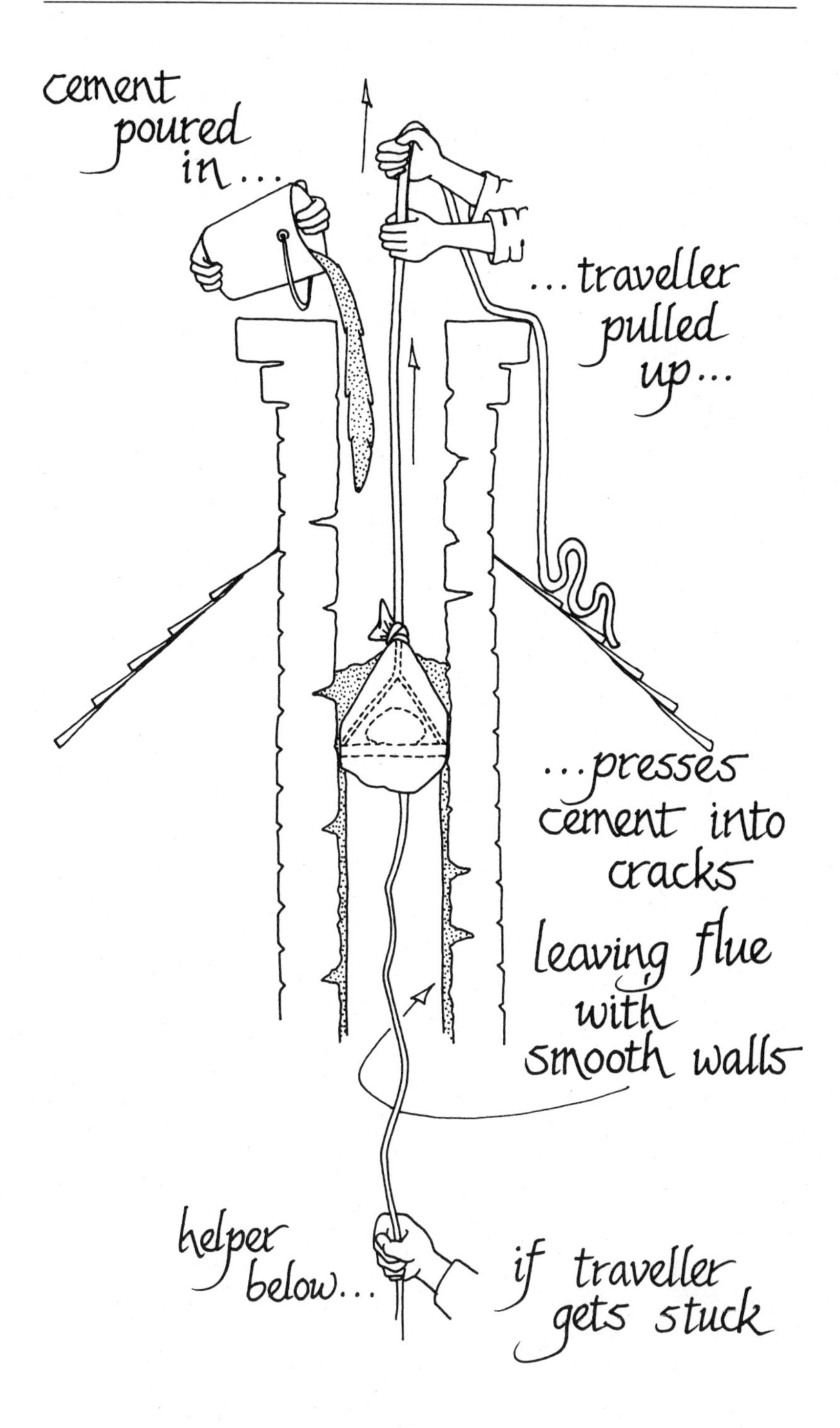
cement poured in…
…traveller pulled up…
…presses cement into cracks
leaving flue with smooth walls
helper below…
if traveller gets stuck

it single-handed and spent half my time running up and down ladders in an impotent fury. One person below, one on the roof and another to mix and haul cement is better. But the cement is runny enough that it will get into the seams and most of it will stay there. Just go slowly, pulling the traveller up one layer of brick or joint of liner at a time, pour down cement to fill the traveller's moving edges and give the cement plenty of time to move into the cracks. Frustrating perhaps, messy for sure, but it's the only way I know to revive an old flue.

Lining the Flue With Stovepipe

If the chimney is simply too far gone for the traveller and cement treatment, you may be able to keep it functional by lining with stovepipe. Most any chimney this deteriorated will surely be old and of sufficient width and of archaic design that you will have no dampers or odd angles in the flue to complicate the work overmuch. Simplest is if you have a big old chimney and want to install a stove small enough that a pipe elbow will go down the chimney. On the roof, assemble your pipe, elbow on the bottom. With a drill or steel punch, put holes through both pipe lengths on both sides of each connection. Then fasten together with sheet metal screws. Needless to say, size of hole should be a thread's width smaller than the screw.

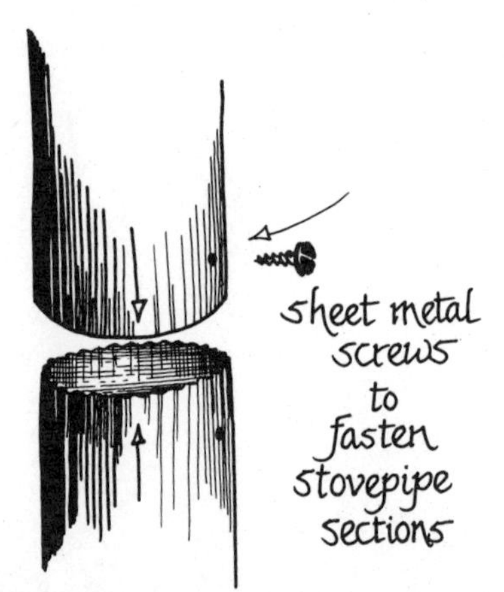

Now, lower pipe down, fastening as you go till the elbow is opposite the opening in the flue. It will take a bit of doing to attach a length of pipe to the elbow through a small opening. You may have to knock a few bricks out. Or perhaps you can cut the pipe to a one or one-and-a-half-foot length. Then one person can

put an arm through the pipe and on into the flue to grab and hold the elbow while another works the pipe length on around the partner's arm. Be careful with this and wear stout gloves. Sheet metal can cut, especially if you run a hand along a newly sheared edge.

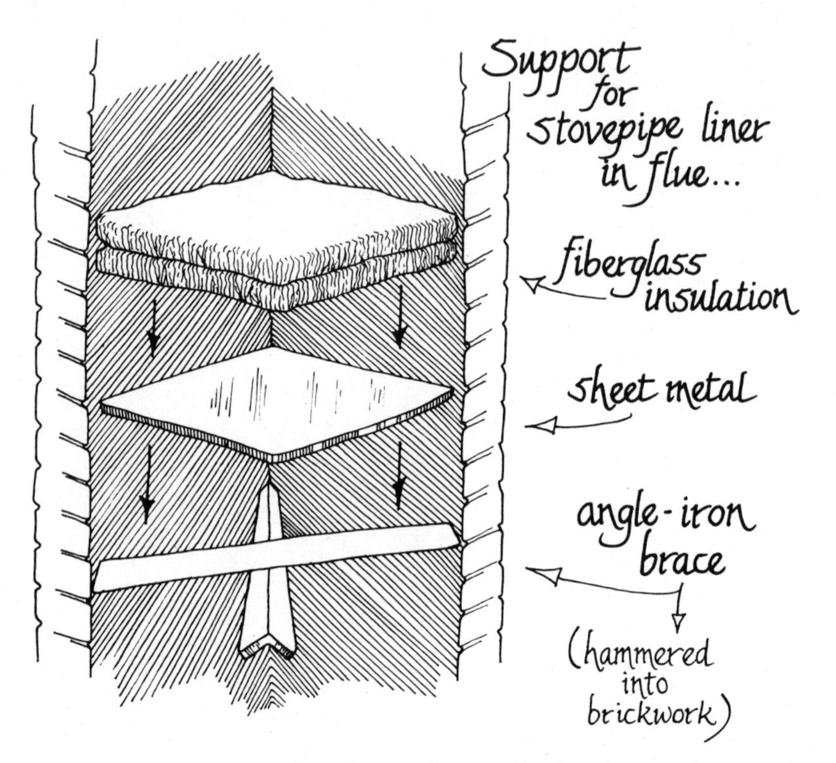

You can support the pipe at the top by simply running a rod through holes in each side, the rod ends resting on the flue top. More complicated, but more permanent would be to provide full support. I'd get heavy pipe to begin with as it will last longer. Then at the opening down in the house, by trial and error, perhaps by snipping ends off a stick, determine the diagonal measurement of the flue. Now get a couple of lengths of angle iron or metal pipe just a bit longer than the two measurements. Use a hammer to wedge the metal into the brick in an **X**, rods touching in the middle, the **X** about an inch below the level of the bottom of the hole. Measure the inside flue dimensions and cut a piece of sheet metal just a bit smaller all around. Roll it up, put it into the flue, unroll it and lay it on the **X**. Now put several layers of fiberglass insulation on top of this platform. Install the pipe as above.

Back on the roof wiggle the pipe so it is as near as you can get it to dead center in the flue. Around it pour in just about any loose, lightweight *fireproof* material you have. Loose rock wool insulation would work if it will fall freely down the flue. I would recommend one of the mineral garden soil conditioners, such as vermiculite. A cubic yard weighs practically nothing, it is cheap in bulk and will not mat down over time. Just test it well to be sure it is fireproof.

You will want to seal the top of the new flue against water. Put a rain-cap or downwind elbow at the top of the pipe. Then, with several layers of fine-mesh poultry netting, build a mesh topping over your inner core. Make it higher by an inch or two at the pipe. Put on enough mesh that it will support your hammer without bending. There should be no holes through it more than an eighth of an inch or so; three to six overlapping layers would do, depending on mesh size. Now, just mix up mortar and plaster the cap so all wire and the top bricks of the flue are covered well by a good half-an-inch of mortar. It's best to apply in two stages as with the stuccoing job we did on the outside of a defective

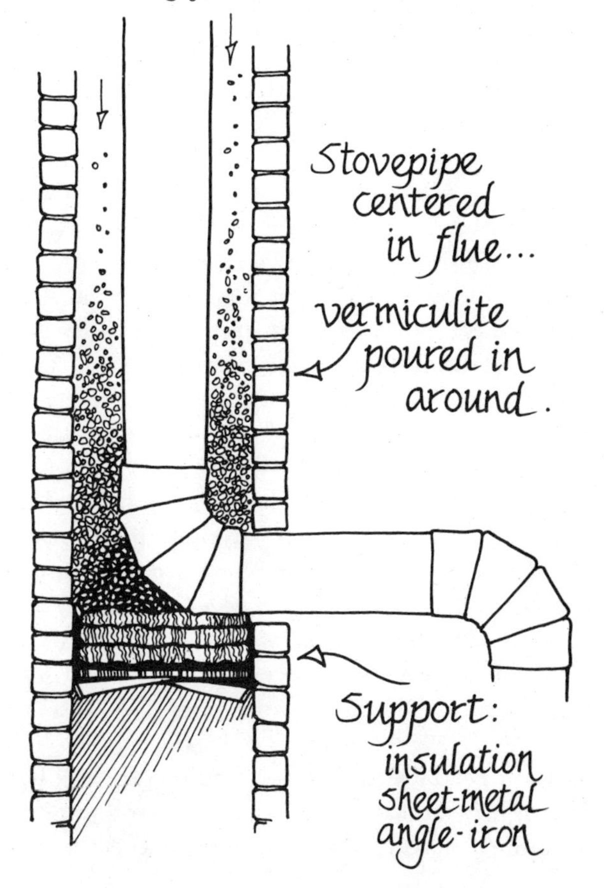

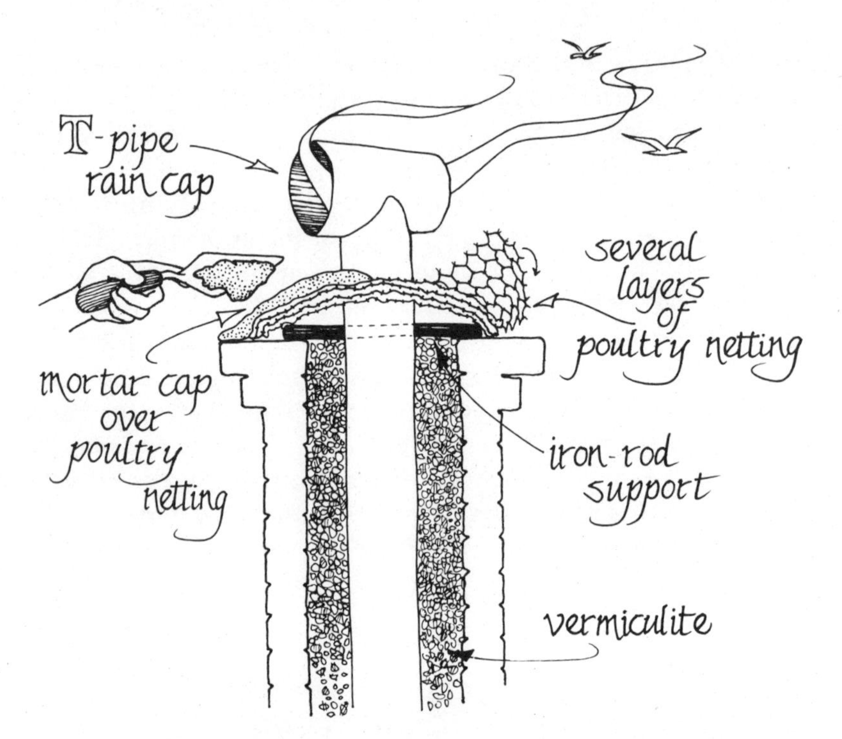

flue—the first layer to cover the wire, the finish coat going on about a half-day later. While you're at it, you may want to stucco up the outside of this chimney. If you do, and the old chimney is not about to topple, it should last a good many years.

Lining a flue over an existing fireplace is more of a job. If you plan to use the fireplace, don't even attempt it unless the fireplace and smoke chamber above it (see chapter 4) are in servicable condition and you can get to the bottom of the flue (where it narrows at the top of the smoke box). You will want to put down the biggest diameter pipe you can, and seal the rest of the flue at the pipe bottom. I've seen some situations where the pipe can be lowered into place, then two people with long sticks, one on the roof, one in the fireplace, could ram fiberglass insulation in around the outside of the pipe and pack it well enough at the base to form a decent seal and hold up the vermiculite outer core. The rest of the operation is as described above.

If the fireplace lacks a damper, you can open a hole in the flue big enough to work through and install a damper (see section on installing stoves). Place the pipe so the damper can be operated through the hole and you've a solution to the problem of installing a damper in a very old fireplace.

It would likely be on the second floor, thus a minor inconvenience, but this is the quickest and cheapest way to control draft in an old flue that you have to reline anyway.

Building Permanent Chimneys

There are only two kinds of long-life flues I'd recommend you try putting up yourself, and I'll admit that you ought to hire professionals to do them—a prefab metal one and a cement block one. Laying a brick or cement block chimney is a task for a mason, one with a lot of experience. Oh, anyone with a fair eye and a mason's level can raise a chimney that will hold up; you see plenty of home-builts out in the country, and, I've built a few.

But anywhere that appearance counts, the unevenness that an amateur is bound to build into the courses will produce an eyesore; it's a fact that bricks and blocks are so shaped that they have to be laid plumb and true overall course after course to look good. Then too, there's the scaffolding to put up and all that masonry and mortar to lug up the ladder. Every mason comes accompanied by one or two really strong assistants. Indeed, I'd not recommend that most people undertake any mortared chimney. Blocks or bricks get awfully heavy up at second floor level, and if one falls, it can cause a lot of damage.

Masonry Chimneys

However, if you must, here's how to put a flue up. You'll need a lot of mortar mix, a wheelbarrow to mix it in, a hoe to mix it with, and water of course (which will freeze before it sets

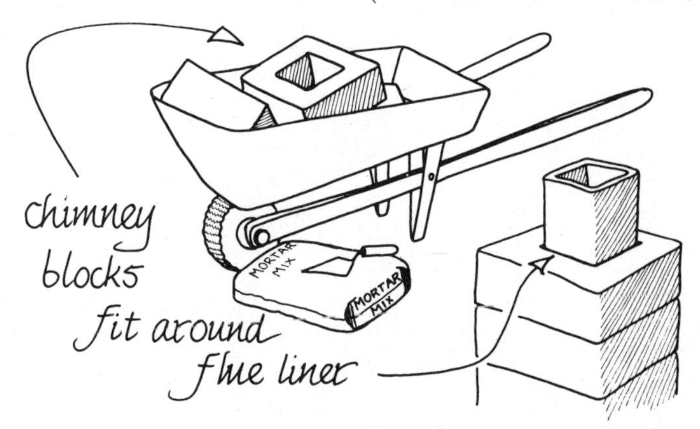

in cold weather, so don't wait till late fall to begin building). The chimney blocks are hollow and come in a variety of sizes; you'll need some solid ones, too, plus clay liner sections also available in a variety of lengths and sizes. The liner should be a bit larger than your stove smoke hole, the interior hole in the blocks just a bit larger than the liner. Scaffolding you can rent from some paint stores or most any equipment rental agency. Ladders too.

Get a pair of steel-toed boots in case you drop a block, gloves, a sturdy spade to dig with, a mason's trowel and a bucket for the mortar. If you have to go up a wood-sided wall or any other flammable wall, get some heavy galvanized wire or strap iron to make keepers with. You'll have to build the chimney out an essential two inches from the wood, and will need some sort of metal attachment between flue and house wall every few feet. Next, get a mason's chisel, a small high-temper chisel with an inch-wide point, and a full set of carpenter's tools, if, as is usual, you will be carrying the flue up through the eaves of your roof. Finally, double check your life insurance coverage and double it if the house is over one story high and you aren't used to working at considerable heights. At this point I might add that a mason can put up the chimney in two days for not much of a fee.

Run your stovepipe out through the house wall. In frame houses, plan to use the same fire-stop box installation covered in detail in the final section of this chapter, including cutting out the wall, using triple-insulated pipe and an insulated, galvanized metal box. The inner shell of the triple pipe should be a ceramic or cast-iron thimble long enough to run from inside the room out to the liner of the flue; this will vary from house to house.

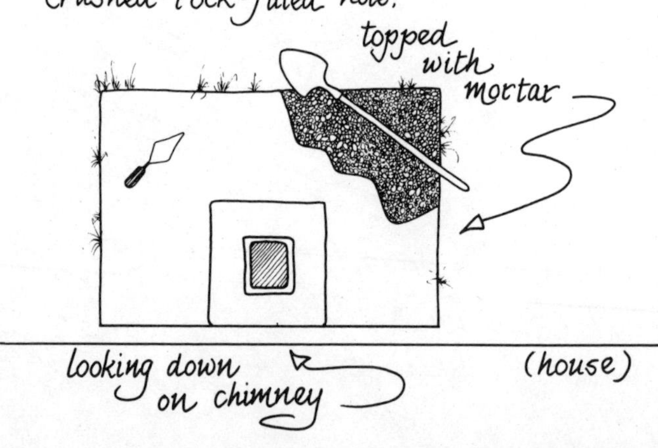

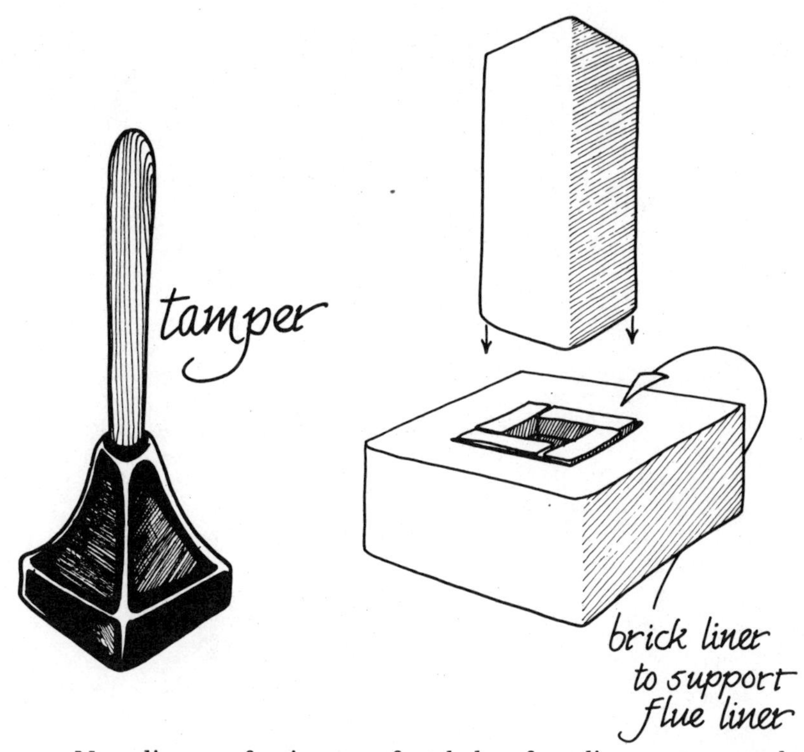

Now dig your footing to a foot below frost line or more and three times the width and twice the depth, front to back, of the outside dimension of the chimney blocks. Pour in two-inch layers of crushed rock and tamp them down till they are perfectly firm. A big sledge is a good tamper, but you can buy or rent a regular tamper, a heavy, square iron head on a wooden handle. Add rock two inches at a time, tamping well till you have a bed at least a foot thick. Now, pour in a good bed of mortar and lay in your first block, a solid one, center plumb below the center of the smoke hole. Butter on a half-inch of mortar and lay on the next, and on up to a couple of feet below the pipe. Each block must be perfectly plumb, square and seated over the other below. Continue to use a half-inch of mortar.

Two feet below the smoke hole build in your cleanout. Use a chisel to score all around one side of the hole in a hollow block, or two, knock out the middle and put the block on with the opening aiming away from the prevailing wind. In the hole you can put a temporary closure of bricks you will have to dig out when clean-out time comes along, or you can mortar in a clean-out door, available at mason's supply houses. If you can get

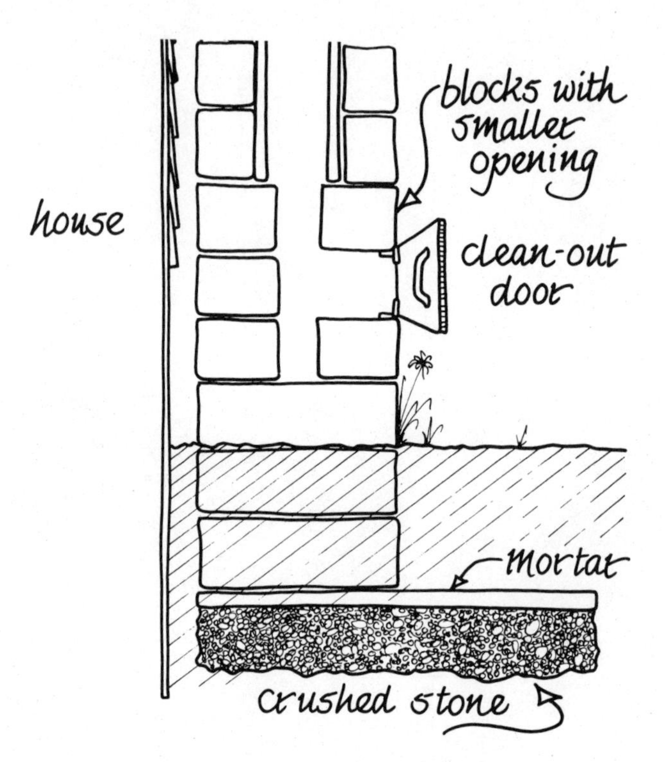

them, the blocks for the cleanout should have an interior opening a few inches smaller all around than the upper blocks in which the liner will go. Put one more small-holed block over the cleanout and you have a solid base for the liner to rest on. If you can't get such blocks, you'll have to build up the liner support with a layer of bricks, mortared inside the hollow blocks.

Use fireclay cement to bind the liners. Butter the base block or bricks with it, set in a section of liner, making sure it is plumb and square. Scrape out the excess from the inside. Put blocks on over and around the liner, keep adding liner first, then blocks. Don't put mortar in the space between liner and blocks. When you come to the smoke hole, use the chisels to chip properly sized holes in blocks and liner, and cement in the thimble. Mortar in keepers every six to ten feet.

When you get to the roof, you're on your own (unless you use stovepipe as illustrated). There's all sorts of stuff up there you'll have to saw into: rafters and facia and scotia and storm gutters and roof sheathing and shingles. Good luck on any sort of per-

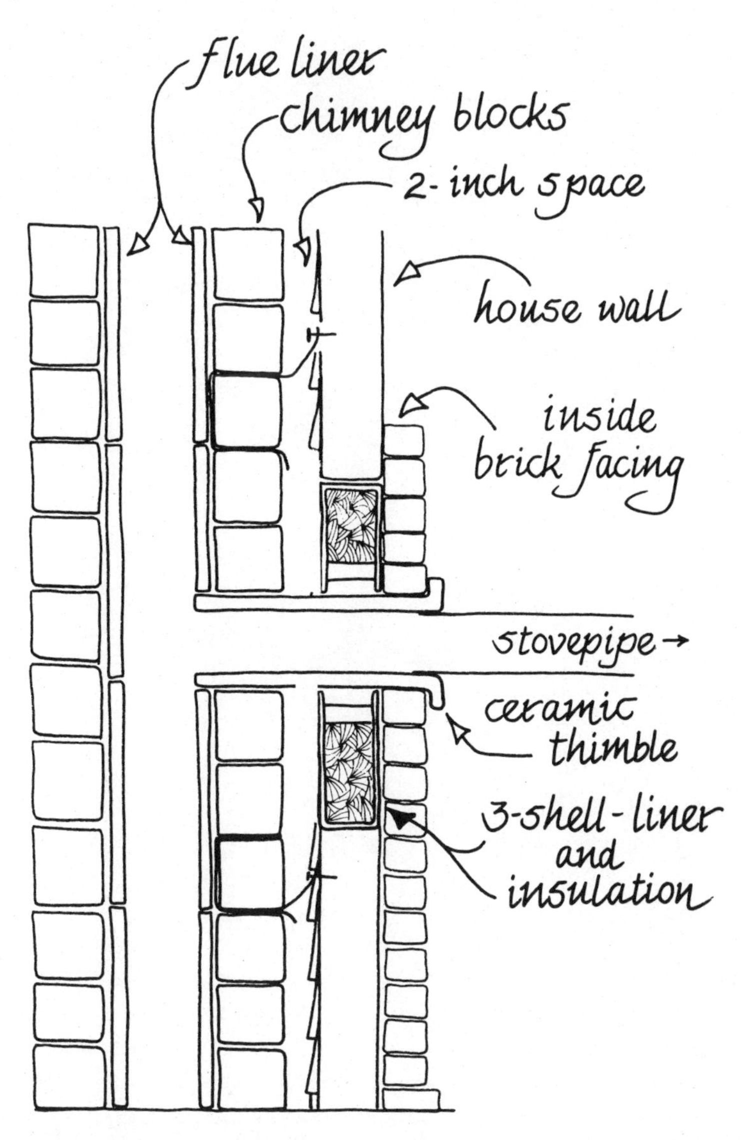

manent residence-type house; I'd call in a carpenter. Just remember to retain that two-inch clearance between flammable material and the flue. Bind in flashing as described earlier and build the chimney to the proper height. Cap it well and put on bird screens, wind deflectors or whatever else is needed. If the space between house and flue bothers you, nail on lengths of overlapping flashing along each side. Let it all cure for several days, then light up.

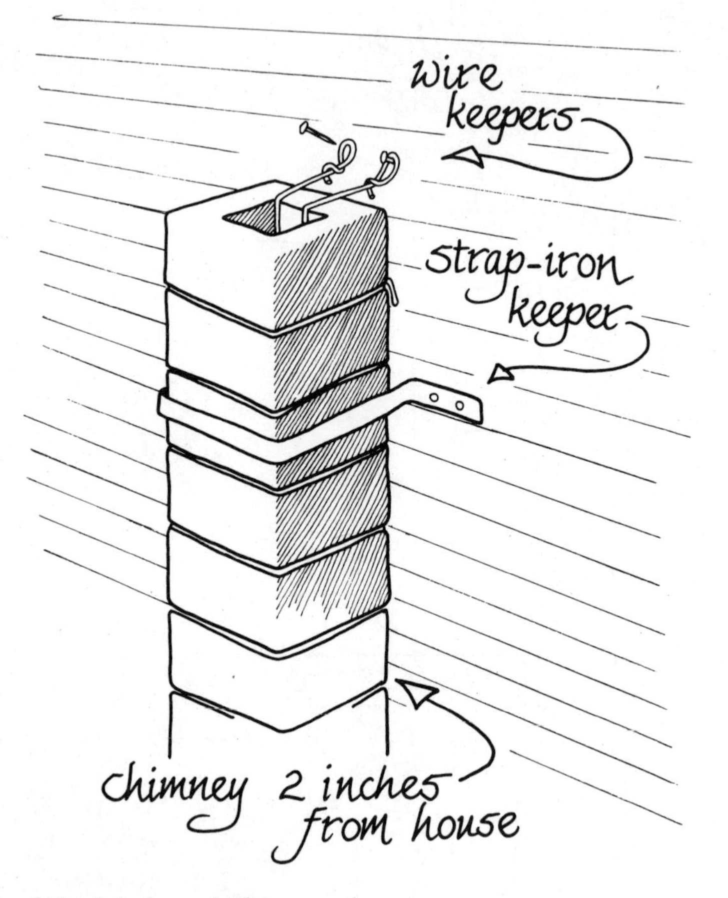

Metal Prefabricated Flues

Easier, but more expensive (at a dollar-an-inch) are the metal, insulated prefab chimneys sold by most heating contractors and such big retailers as Sears Roebuck. Metalbestos and similar brands enclose a fireproof insulating layer between two metal tubes. The flues are safe so long as curves are kept to a minimum and you retain the two-inch clearance between the outside and any flammable material. Many installers stuff fiberglass insulation in around much of their pipe where it goes through the floors and around wood as an added protection. Each brand is different, going together in a different way and needing different accessories such as fire breaks where they go through floors and caps where they come out through the roof. All I've seen come with excellent instructions if you want to do it yourself. A heating contractor and his helper have all the drills, saws and other tools needed and can do the job for you in a day. All I can add is to plan the run of pipe and stove location carefully. Remember, you must keep that two inches all around throughout the entire length. You will probably want to avoid cutting through major supporting timbers in floor or roof, so plan to locate the fireplace or stove accordingly. If you aren't very familiar with house construction, I'd not recommend that you install even these flues yourself. For example, modern houses no longer have electric wiring encased in the old steel spiral guards, and a drill being used to get a start on a hole to be sawed in a ceiling could get you electrocuted if the bit chews through the plastic-bound conduit. The illustration will give you an idea of the work involved in putting a prefab up, and if you think you can do it, my best to you.

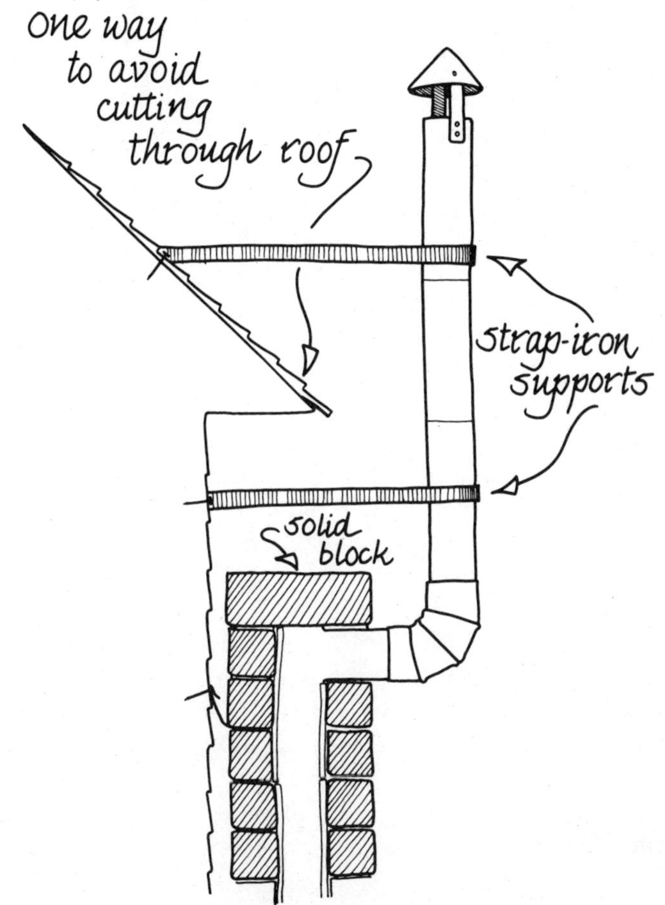

There is also a triple-wall variety of prefab pipe on the market that becomes barely lukewarm to the touch under the hottest fire and doesn't need the two-inch clearance. The twin-layer plus asbestos flue will run you that dollar-an-inch and the three-layer pipe can run higher. With either you are paying a lot of money to a heating contractor or store and even if you don't have those folks install the chimney for you, you should be able to tap them for all the advice you need and maybe even borrow a few tools over the weekend.

Flues for Temporary or Intermittent Use

You may encounter flues that consist of nothing but stovepipe, or in rare cases old "cattied" chimneys of mud and sticks. Neither are suited for long and constant use, but do fine where heat is needed infrequently enough that a major investment in masonry or prefab chimneys is not justified. We'll build one of each from scratch, which is pretty much what you will have to do if an existing one is in poor repair or improperly installed (usually the case, especially with stovepipe flues).

Stovepipe Flues

All the books and reference guides tell you never to build your flue of common stovepipe and few, if any, building codes in populated areas where building codes serve a needful purpose will permit it. And I go along if you're contemplating anything approaching a permanent installation. Stovepipe is flimsy stuff and liable to all sorts of problems under sustained use. But for your two-room log cabin out in the deer woods that receives no more than a few weeks cold-weather use each season, or your shop in the garage that's heated only now and then, a stovepipe flue is just fine—so long as you take proper precautions.

First, plan at the outset to dismantle and clean out a stovepipe flue after every period of use. That way you'll automatically avoid buildup of potentially flammable soot and creosote. Besides, if you leave the pipe up during several months' disuse, the weather will rust it out faster than a continuous fire. If you'll be using the flue frequently enough that dismantling between uses is too much bother, you will probably be using the place and the stove too frequently to make a stovepipe flue both safe and practical. Build a permanent chimney.

If you can get by with a quick and easy intermittent flue, however, stovepipe can serve. There are two steps in putting one up. (We'll get into installing stoves and building fireplaces later on.) First, you must guide the pipe safely out of the house, through wall or roof. And second, the pipe must be supported and guyed well enough that it will stay up no matter what the weather. If any significant amount of pipe is retained inside the house it must also be secured well enough that it will not fly apart in event of a flue fire.

In the old days great lengths of pipe were often run across whole rooms before entering the chimney—the pipe often radiated as much heat as the stove. Pipe flues for more modern oil-fueled space heaters used to be run out through a simple sheet metal plate with a hole in it that was nailed over a foot-square hole in the building. Plenty of old places still come equipped this way. However, such an installation is dangerous with a wood stove. Unlike kerosene, oil or gas space heaters, fireplaces and wood stoves release great quantities of heat into the flue. And unless you run the pipe inside for many feet—allowing the hot gasses to radiate a lot of the heat before exiting from the room, the sheet metal plate can heat up enough to kindle the adjoining woodwork wall, lath or wallpaper.

The pipe should exit a frame dwelling through a fire-stop box we mentioned in the cement-block chimney section. You can purchase complete units from heating contractors who sell the metal, insulated prefabricated chimneys. But for a cheap, easily transported installation for a remote or seldom used location, you can build your own. You'll need materials for a box in each wall or floor or roof you plan to go through. Most folks take the easiest route out through a sidewall or a window. Just be sure to follow the basic principles, which the building industry has incorporated into the prefab fire-stop boxes. Basically, you want the hot pipe surrounded by two circles of dead air, so you make a "three-wall" pipe that will be cool to the touch no matter how hot the inner pipe gets.

Materials for the Stovepipe Flue

First, you'll need two squares of galvanized sheet metal obtainable from any heating contractor. Get the thickest they have for durability's sake, so long as it is not too thick to be cut easily

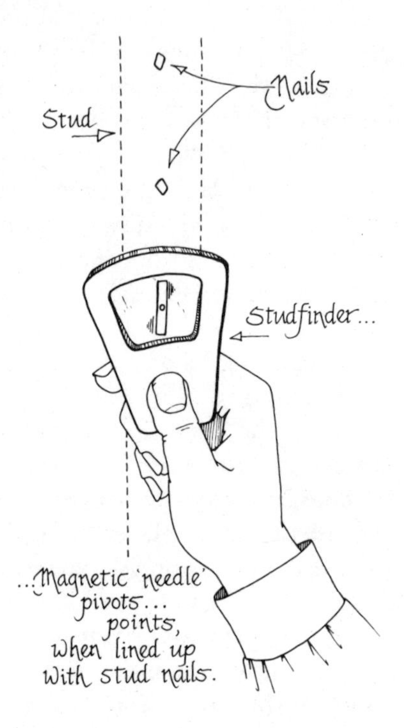

with tin snips. The sheets should be three or four inches wider than the distance between supporting timbers in your building (called joists in a floor, studs in a wall, rafters in the roof). You can determine this in plastered houses with a stud-finder, a cheap little magnet obtainable from any hardware store. It indicates location of the nails holding floor or wall covering on the framing members. Most frame structures have studs 16 or so inches apart, so the sheets should be twenty-inches-square. You may find it easier to put the pipe through an existing opening, the top sash of a window, for example. Just get enough sheet metal to cover the opening with enough overlap all around to fasten it to the frame. In addition you'll need a good set of tin snips, hammer and nails, saws and chisels to cut through the walls, two lengths of lumber the same size as the framing (usually 2 x 4) and the correct size to fit horizontally between two studs plus some sheet metal pipe and accessories you'll find at most hardware stores.

Your stove will have a smoke outlet sized to fit one of the standard pipe diameters: six, seven, and eight inches most likely.

For the box get one length of the stove's size and each of the two next larger sizes. For the usual small stove with a six-inch outlet, that would be a six, seven and eight. Pipe comes in two-foot lengths and is usually sold unfastened (open along the seam). This way sections can be nested together and are easy to transport. You'll also want a small bag of loose fireproof material as we used in lining the old flue with stovepipe. And finally, get two collars—donut-shaped fittings in the stove's pipe size. A storm collar would be best for outside and decorative collar for inside, but two of one or the other will suffice.

Construction

At the site, cut your way through the wall, ceiling or whatever. In brick or stone construction, the opening need be no larger than the stovepipe. Just be sure that no margin of the opening is closer than 18 inches from any woodwork. And, if in going through stone or brick you find a wood inner structure, make a hole as for frame construction. Again keep that 18 inches clearance away from anything combustible.

In a frame building, do your best to cut a square hole 36 inches on a side (or a round one with a radius of 18 inches) through the ceiling, roof or wall. No need to try to cut through the main framing members, unless they are so close together that you won't have a good two inches of space on each side of the multiple-wall pipe we're going to install. Most modern buildings are made with the vertical studs spaced 16 inches from center of one to the other. Most old houses, camps, shacks, barns and places where you'll be using this sort of flue, were put together the way the builder felt like doing it at the time. Log cabins don't have any studding. Last place I visited with a stovepipe flue was an old steel Toonerville trolley hauled to the edge of a lake for a fishing cabin. Good fishing, too. What I'm trying to say is that we can't give full instructions on how to make your opening. Just do it as best you can, and make it as close to the good and safe 18 inches clearance as you can. Drill yourself a hole in as many thicknesses of partitions as you have to go through, and hacksaw metal, use a keyhole saw on wood, chisel on stucco or interior plaster and your chain saw (pushing in gingerly with the end to get started) in log construction. The hole should be directly above or behind the stove location if at all possible (for details of

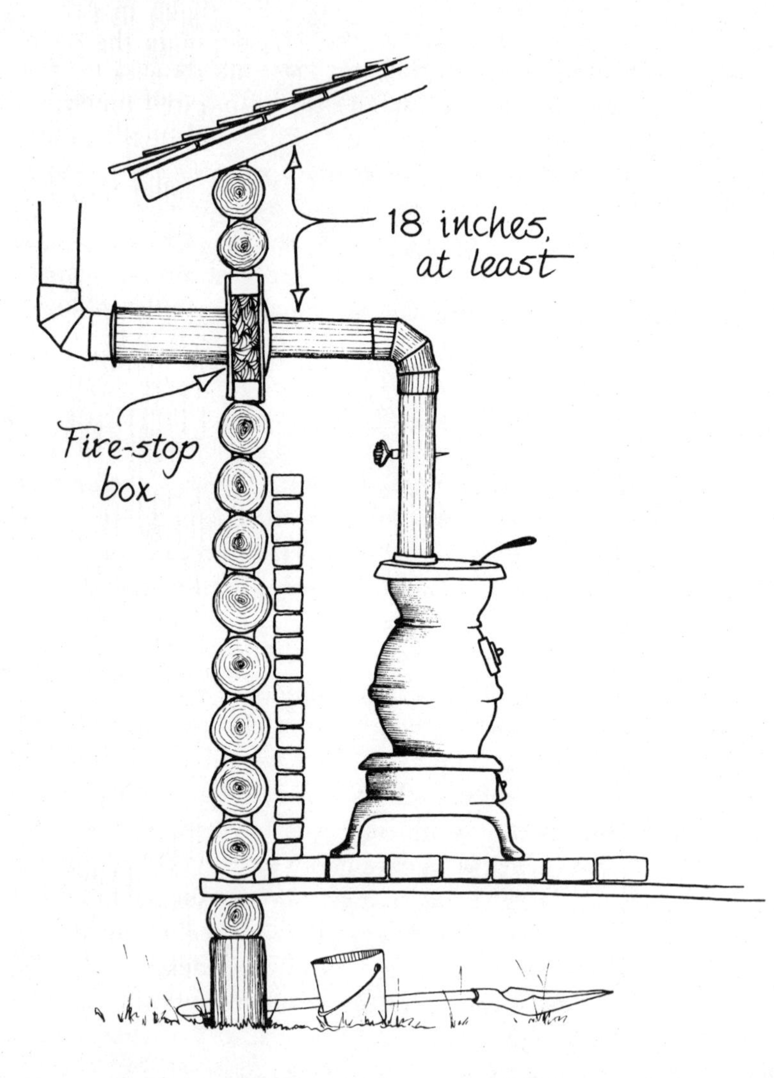

stove location, see chapter 3). But if for some reason you must angle the pipe, try to have no more than a 45 degree slant. If you want to vent out horizontally behind the stove, the center of the hole should be right in back of the stove's smoke outlet (height of center of smoke hole from floor plus height of whatever hearth you are using—see section on stove installation once again). It is better economy to keep as much heat-radiating pipe as possible inside the dwelling, so exit the pipe near the ceiling. Just be sure it comes no closer than that eighteen inches to any combustible material, and that includes the rafters of your ceiling.

The Fire-Stop Box

Now assemble the three-part fire-stop pipe. You'll notice that one edge of the seam has a groove stamped in it, the other a flange. Fit them together at one end and work up along the joint, pushing down as you go. It's easy to do this way. If you tried to mate the edges along the entire length in one operation you'd go nuts trying. Put the middle pipe into the outer one and stuff in the insulation good and tight, so the inner pipe is centered. Then do the same with the smallest pipe. Assemble them with crimped ends of all pipes at the same end.

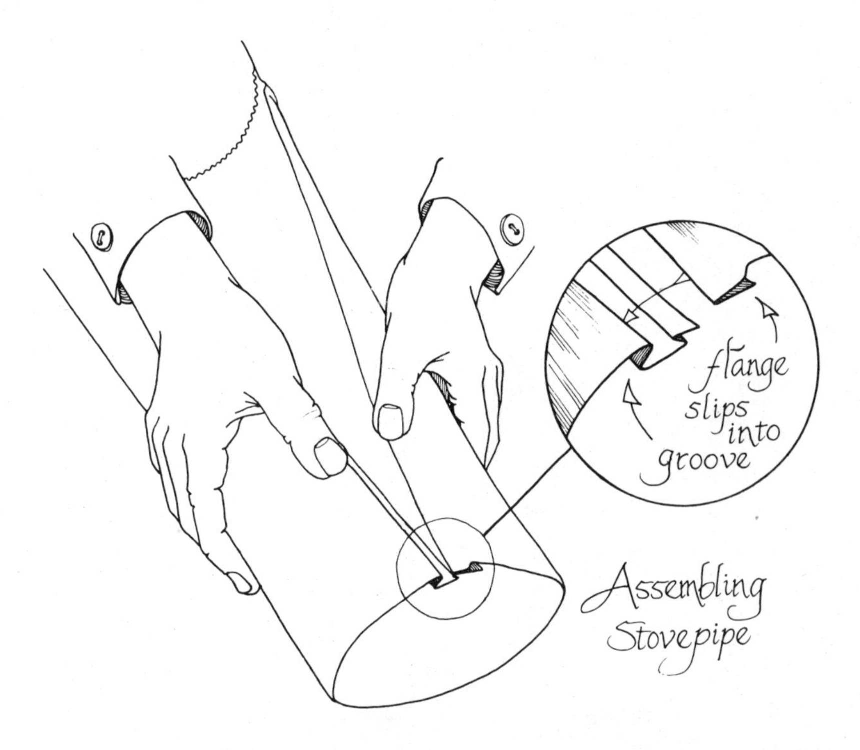

Now, provide a good support on all sides for the metal sheet, which will hold in your fire-stop box. The main type of construction where you'll use this apparatus will be traditional frame. You've got to build a square box, so you must make horizontal members to fit between vertical studding in walls, intrajoist or rafter members in floors or ceilings. It's probably easiest to saw them just a bit larger than the space they are to occupy and hammer them in, to be held by simple tension. You may want to nail them, and often there are several inches of lath, plaster and air to go through. Use long nails, and good luck. Or, maybe your walls will be sound enough to hold the (comparatively light) weight on their own.

Now, cut holes to fit the three-part pipe into the two sheets of galvanized; slip the assembled three-part pipe into it, with the crimped ends on the *outside*. Support the pipe horizontally and pack more insulation into the square hole in the wall, bringing it out even with the inside wall. Then slip the other piece of sheet metal over the pipe and nail it on. The pipe will be two-feet-long and you can adjust it any way you want in the wall. For appearance's sake, you may want the inner end flush with the wall so only one size pipe shows in the room. Put a collar over the pipe where it enters the wall and the end of the three-part pipe will not show. And you have a good, safe exit for the stove. Another collar inside will hide the insulation and make for a tidy appearance.

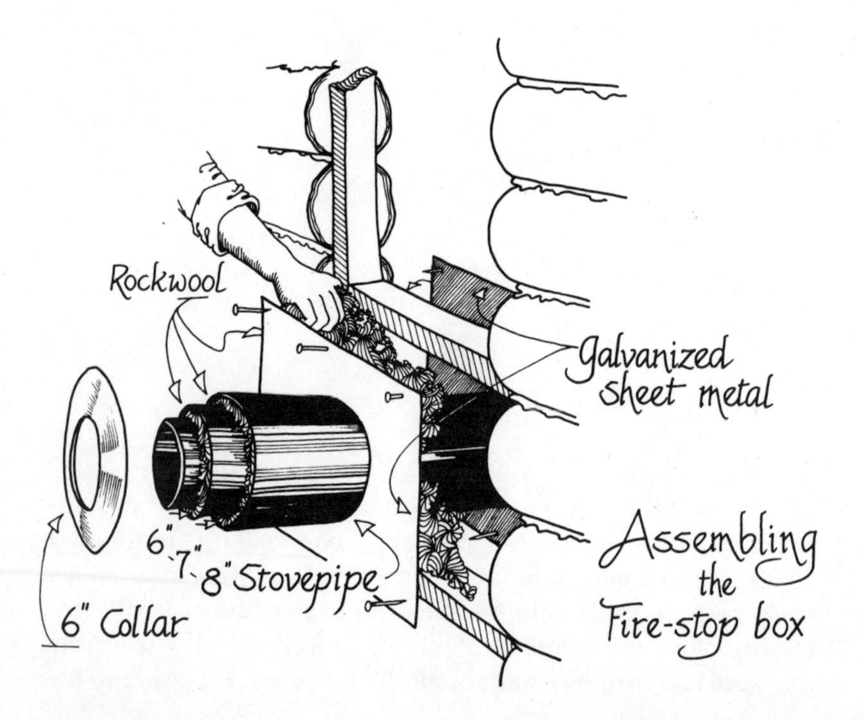

Outside

There will be a foot or so of three-part pipe showing outside the house. This may be enough to clear the eaves of your roof. Or you may have to cut a small piece of stovepipe to extend it. There are two ways to support the vertical run of pipe. For short flues typical of one-story cabin installations, just put on an elbow (a four-piece assembly adjustable in angle) and enough pipe to rise at least two feet above the eaves. You can support the elbow with a length of wire running down from the eaves on each side. Another length of stiff wire—a couple of coat hangers—can be bent around the top of the pipe and nailed to the roof or eaves.

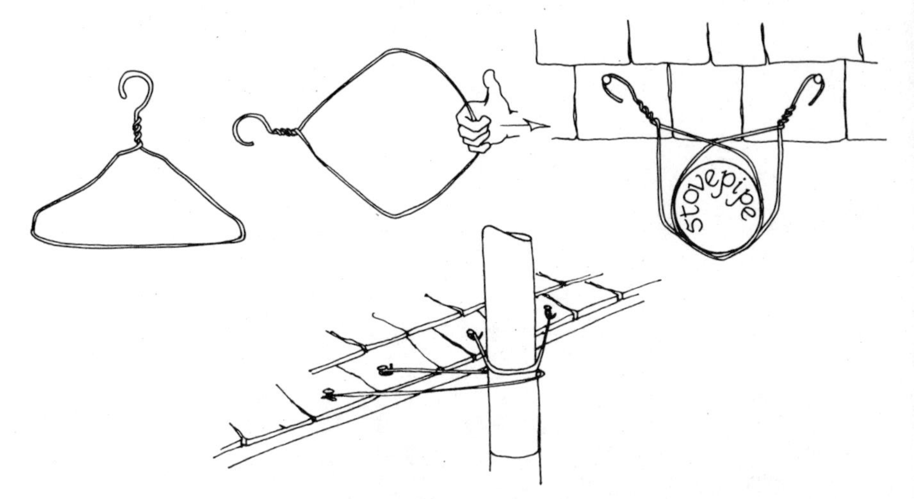

Better—safer and sturdier—is to support the flue with a wall bracket and roof-edge metal bracket. You can purchase wall brackets from a heating contractor or you can make one or have any welding shop put one together for a few dollars. The roof-edge bracket you'll probably make on-site from a length of galvanized sheet metal. If the pipe runs, say four inches from the roof, take yard-long pieces of foot-wide sheet metal and bend it in the middle into a **V**-shape. Cut holes a bit bigger than the pipe near the bend in each half, run the pipe through and nail the ends to roof and eaves.

At the bottom, put a **T** on the end of the inner pipe of the three-part assembly with the other collar placed to seal the outer end—to keep water and critters out of the stuffing. Attach the shelf bracket so the bottom of the **T** rests on it, then put your pipe on top of the **T**. It is a good idea to support the flue at each junction of two pipe lengths, should you have to go up a high wall. Stiff wire looped around the pipe with ends nailed to the wall is fine. At the top put on a rain-cap, still sold in some hardware stores. You can make one yourself easily enough, though, from an old can. If there is a lot of wind, another elbow put up top and aiming downwind may be the best cap. Finally, it is a good idea to fill the bottom of the **T** with sand. This will add a bit of weight to the bottom and increase stability in a wind, but more important will prevent any loss of draft from air getting in between pipe bottom and the shelf.

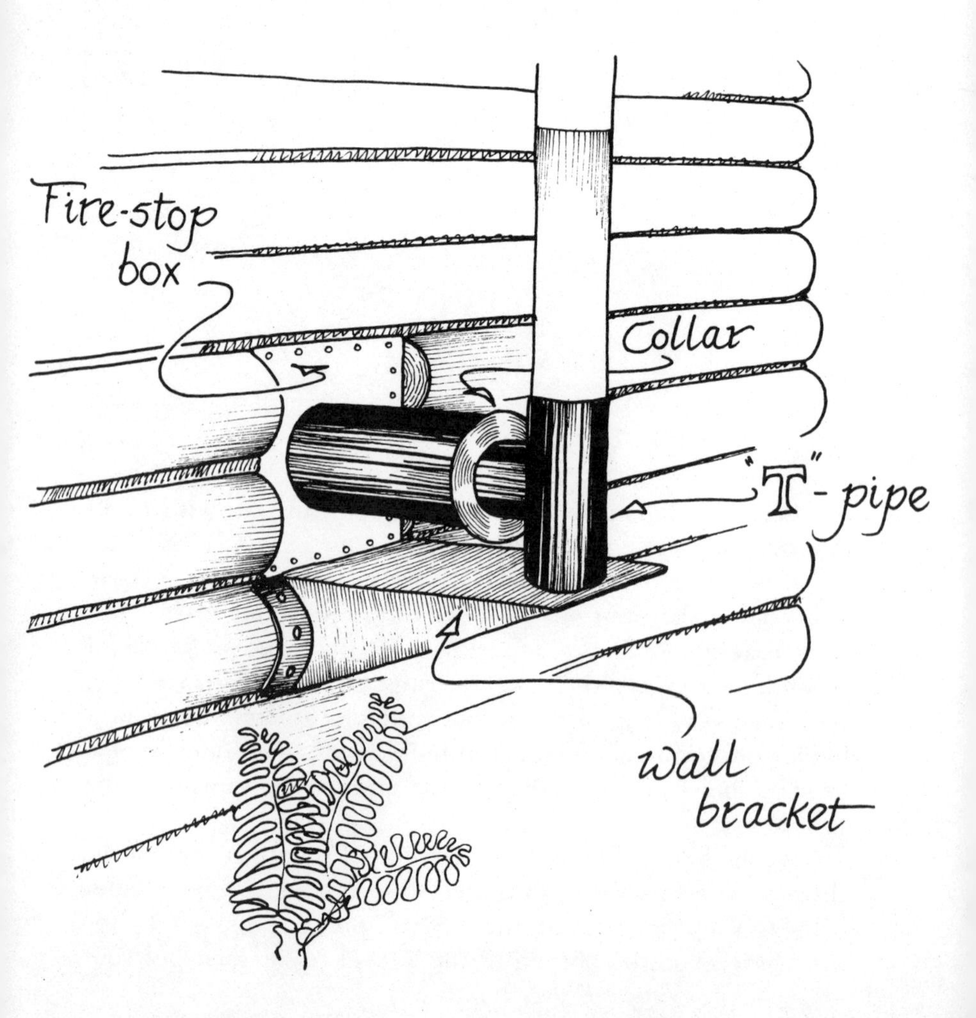

Cutting Galvanized sheet metal with Tinsnips
Galvanized, sheet metal,
Roof-edge bracket...
...attached under shingles, under eave

Make sure that all supports fit loosely, so disassembly will be easy when it's time to leave. Just pull all pipes apart from the bottom and slip the flue down. (Unless it's very tall, you should be able to assemble again when the time comes from the ground up.) Shake out all accumulated soot and put the pipe in out-of-the-weather storage. The only chore remaining is to seal up the opening through the wall. Easiest is to buy an end cap. They have handles that make for easy removal. Grease the end of the pipe, though, else pipe and cap may rust together. Lacking a cap, a coffee can may serve. Or you can just push the pipe inside and tack another piece of metal over the outside hole.

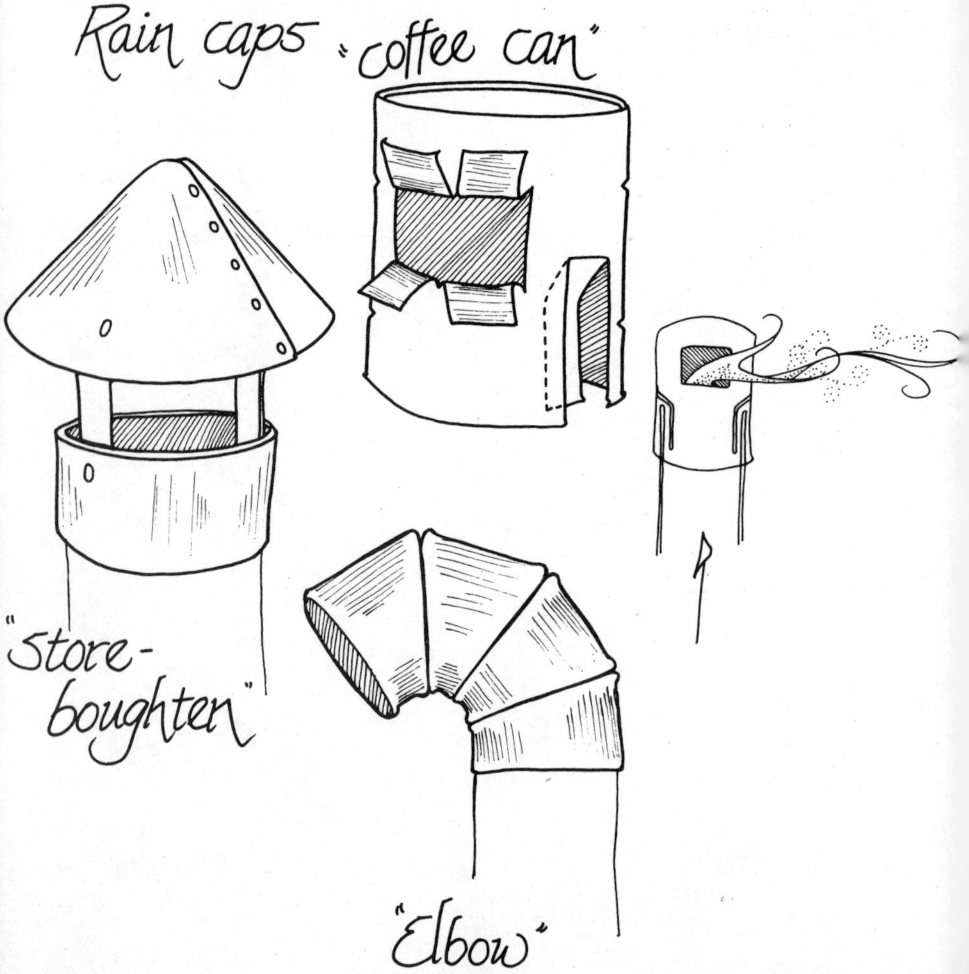

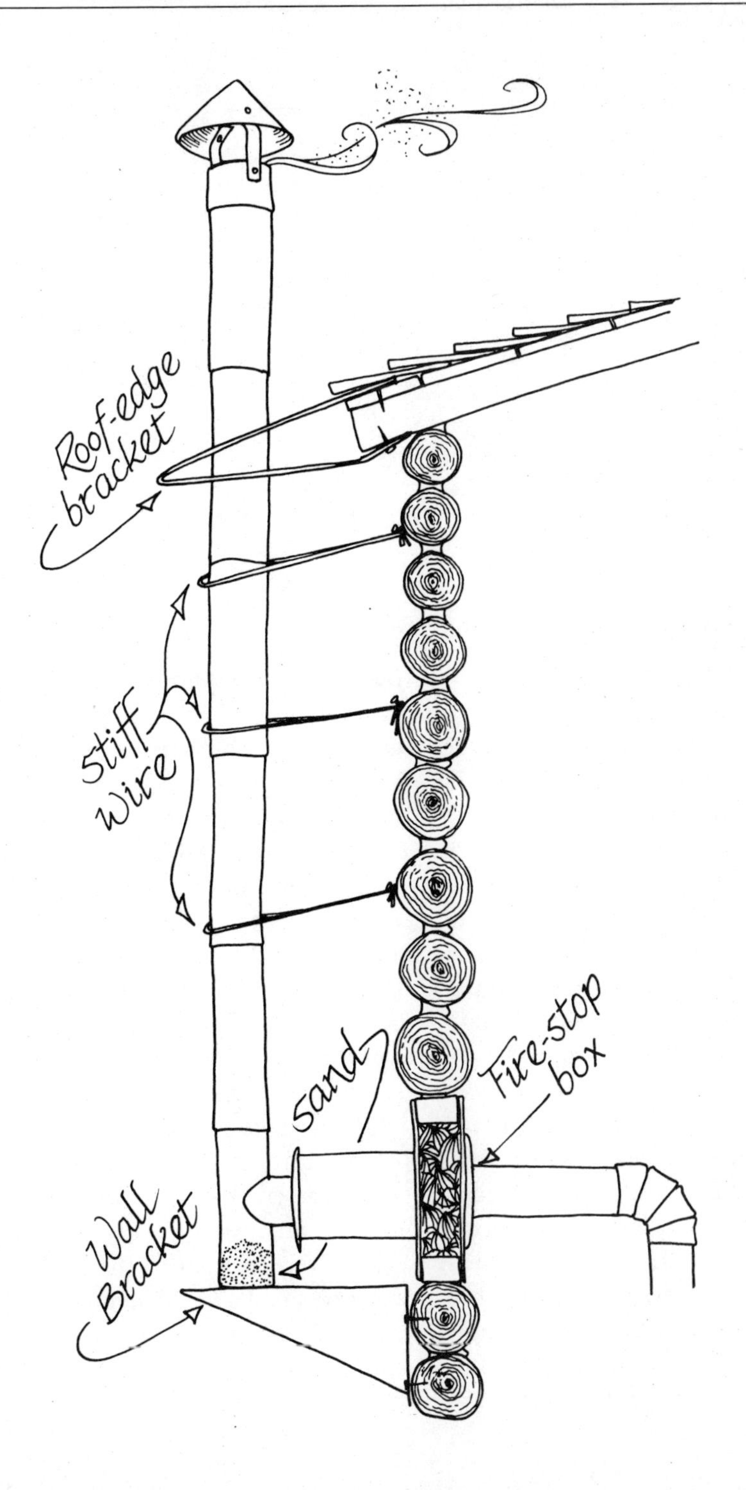
Roof-edge bracket
Stiff wire
sand
Fire-stop box
Wall Bracket

Mud-and-Stick "Cattied" Flue

The very first chimneys built by the colonists in America were of mud and sticks, and you can still find a few original cattied flues in the backwoods of some of the Southern states. They do require a lot of maintenance, and they can catch fire if neglected or overheated. Together with the wood shingle or straw-thatch roofs common during the first years of the colonial period, these "cattied" flues were responsible for so many fires that they were outlawed in many cities. But they are fun to build, will last for a good while and serve safely if kept up. And they cost nothing, require you to pack no more into the cabin site but your own muscles, the food to fuel them and a piece of string.

First job is to find a deposit of the proper mud. A good forest loam won't do. Neither will a sandy soil. What you want is clay, and the finer, gummier and stickier the better. (Some of the best cattied chimneys are found in the state of Georgia—red clay country.) Dig down to subsoil, scout the creek or riverbeds within easy walking distance from the site. Then make a sausage-like roll of each sample of likely mud, let dry and put in the cooking fire. Leave them there through several good fires. Then test. Any that crumble after baking are no good. The ones that harden up are the better choice, the one that is hardest to break and that makes the most musical noise when you tap it is the best of the lot.

Now, pick your flue site. (For either a stove flue or a fireplace; if you're building a fireplace under the flue, follow directions in the fireplace construction section; for a fireplace, I'd try a stone-and-mud Rumford, though I've never made one so I can't tell you how.) Cattied chimneys lack the integral strength of a masonry or stone type. So, you should give them as much support as possible, by running them up the highest wall of the structure they will be heating, the peaked end wall usually. Try to satisfy the basic rule of flue construction and plan to have at least two feet of chimney rising above the nearest line of roof within ten feet laterally. This, so the roof will not interfere with airflow upward out of the chimney.

Constructing the Chimney

Up against the cabin wall dig a hole that is a foot larger all around than the flue base is to be, and be sure to dig down to

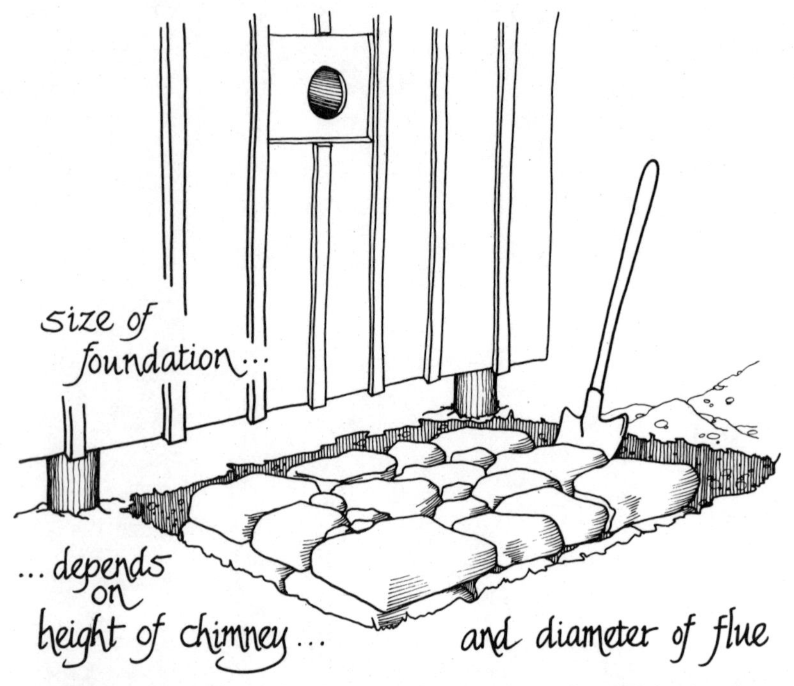

well below frost level in the North, or to good solid bedrock or subsoil in the warmer parts of the country. Fill the hole with rocks if you have them, putting them down in layers, "cementing" each layer with a thick clay and water slurry as you go. If you've the time it is best to let each layer dry for a day or two before adding on the next.

Once the base is down, dry and settled, you can begin building the flue. (If you lack rocks for a foundation, begin the flue at the bottom of the hole.) First, decide how high it is to be—the top some two feet above the ridgepole if you are building near the center of an end wall as recommended—and how big around at the top—three times the size of the flue opening. For a small stove needing a six-inch flue, the top should be 18 inches on a side. For an eight-inch flue, the top should be a full two feet on a side. Now for each foot of height you will want to add on three or four inches in base width. If the top is to be 16-feet high, the base would be four or five feet wider than the top. Say, six-feet wide for a six-inch flue, seven-feet wide for an eight-inch flue. You can add on a foot or two of depth too—extending out from the wall. The added heft at the bottom is to provide stability and a good slope so rain will run off.

Use your string to outline the form of the flue on the building's wall. Tie a rock to one end, tack the other to the roof in the center of the roof opening and score or draw a line along the string. You can also outline the outer sides by tacking the string to the outer limits at top and angling out to the greatest bottom width and marking the line.

If you have stones it is good to lay a stone base up to fire-box floor or the hole where the stovepipe exits your building. But don't build with stone much above shoulder height unless you have mortar and a conventional flue lining—and the need for a permanent, constant-use flue. A mud-and-stone flue could work loose too easily and a wobbly rock pile that high is a real danger to passing deer and other folks.

In stick-and-mud construction, the sticks act mainly as an internal scaffolding to hold the clay in place till it can dry and be baked to hardness by the heat of your fire. You lay the sticks out in an overlapping pattern of squares and pack the mud in around them with your hands, a trowel or small shovel. There are several ways to accomplish this that I know of and I have tried all of

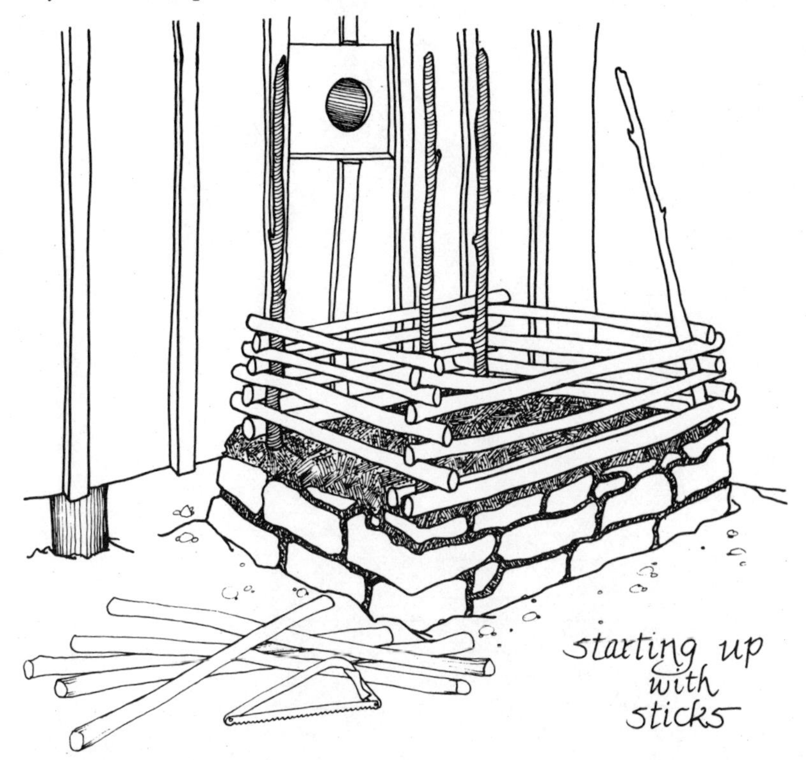

them out at one time or another, and you'll probably come up with some ideas of your own. You can use a single column of interwoven sticks or use two, one of smaller sticks inside the outer one. Just be sure you have a good two inches of mud lining the inside of the flue. If you like, run "corner posts"—straight saplings with twigs trimmed to an inch or so up the corners of the flue, just inside the outer sticks. Just make sure all sticks are green, so they will shrink along with the clay as it and they dry. They should be perhaps an inch in diameter at most, a half-inch at minimum. In time they'll turn to char in the chimney.

Completing the Flue

At most wilderness camps a goodly portion of the food supply usually comes in cans. The only mud-and-stick flue of any account I ever helped build was for a little 15-gallon gas-can stove

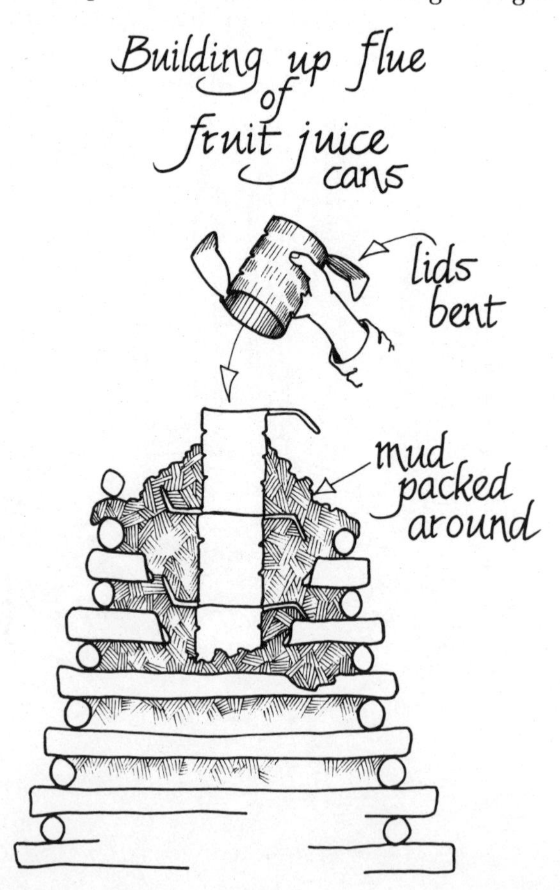

with a four-inch smoke hole. The inner liner of the chimney was made of fruit juice cans of the just-under-two-quart size. The flue went up at the rate of juice consumption, one or two cans a day. The ends were not quite cut out, then bent out and into an **L**-shape, so as to bind in with the mud and sticks. This made for a good, smooth flue lining and reduced need for careful trowelling inside the flue; we just added on a can and packed mud around it. With a flattened can on top during disuse to keep out water, the little flue served as long as we needed it and may be standing yet for all I know.

It is a good idea to use rocks, flattened cans or sheet metal to cap the flue. This will keep water damage to a minimum. And plaster as much mud as you have time for on the outside. Be sure to slope the sides well and dig water channels around the base so rain will run away from the flue.

Going in to a stove flue, leave a hole in the flue that is about twice the size of the stovepipe and locate the pipe in the center. (A large tin can with both ends out is a good tube through the flue side.) The flue will shrink and if you pack the pipe into the masonry tight at first, it will pull the pipe down, crush it or pull it through the wall. Other accessories you might try include wall keepers, to fasten the flue to the building more securely. As you build, mortar in lengths of tin can or stiff wire bent in an **L**-shape. Have the long part well into the chimney, with the shorter part extending horizontally along on either side of the chimney. After the flue is well-dried and shrunken down, staple the keepers to the cabin. Or you can form small circles in the other ends and nail them on.

Burrowing insects—several kinds of small-colony wild bees and wasps in particular—love to dig their spring brood-rearing nests or fall hibernating burrows into nice, warm mud chimneys. In time they'll have your flue spouting smoke from a hundred places. I suppose you *could* mix chlordane or another bug killer in the outer layer of clay, though the downwash will kill everything in the soil for yards around. An annual spring application of plain whitewash will deter them more safely. Or you can lug in several sacks of premixed portland cement. Mix up a slurry of clay and cement (I've never bothered to cement up a cattied chimney, so can't suggest a proper ratio, but would start out with one-quarter cement and the rest clay) and plaster it on.

if it rains before mud has hardened...
chimney should be covered with overlapping layers (...sheets, plastic, brush...) and weighted down...
lest it all wash away...

Be sure to wet the outside of the flue before plastering, though, so the mix will hold.

Fire up a mud-and-stick flue gradually. Build no fires for a week as it dries, then small ones for the next few weeks till all is dried down. If the clay is good quality, you should have few problems with the inner lining which should bake good and hard. Keep the outside plaster in good repair and the flue should last for a good long time.

There are any number of additional embellishments you could add. Build in tubular lengths of chicken wire in place of or in addition to the sticks and the structure would be a good deal stronger. A clean-out hole for a chimney used with a stove could be included in the outside wall—just build in a juice can about a foot below where the stovepipe enters the flue. When the can rusts out it's time to dig out the accumulated soot and the can. Then plug the hole again by plastering in another can. You can increase the flue's life by wrapping it in weatherproofing, tin-can shingles or strips of building paper.

Still, for an intermittent-use flue, I'll stick with stovepipe, leaving the time-consuming mud-and-stick construction for the really backwoods hunting camp, for recreation or just to prove to the Scout pack that a real working flue can be built with materials that nature offers gratis—just like the wood used as fuel.

CHAPTER THREE
Heating With Wood Stoves

As Ben Franklin realized, a stove sitting out in the room, surrounding the entire fire and radiating warmth from front, sides and back is the most efficient heating device to put at the bottom of your flue so we'll do that first. If you've enough years on you, you'll recall the good stove installation featured on Norman Rockwell's covers on the old *Saturday Evening Post* (real Americana, if not yet acknowledged as fine art). At least once a winter a cover would feature a batch of Yankee characters trading tall tales around the big potbellied stove in a New England general store. The few old-time general stores that haven't been wiped out by supermarkets relegated the stove to the back of the building when they put in central heating. But in the old days, the stove with its nickel-plated footrests and towering smoke pipe occupied the place of honor, smack in the center of the store. An odd assortment of chairs, a checkerboard on the pickle barrel, a brass cuspidor at the back near the stove for those who couldn't get an angle on the open fire door plus a fair sampling of the retired farmers from the township completed the scenario. But the central figure was that stove, pouring out heat, beckoning companionship, drying mittens and providing a crackling fire to ponder during lulls in the conversation.

Central heating units may be more efficient and bother-free, fireplaces more aesthetically pleasing, but nothing on earth heats better than a black-iron wood stove—except perhaps the sun on a fine summer day. The stove fairly radiates good cheer as it warms up your outside, and the hot water in the teakettle on top is right there 24 hours a day, waiting to warm up your insides. If you're just in from a hike in the snow, put your boots right up on the footrest so they almost touch the stove, though they'd best be unlaced first. Wet soles can overheat and stew your toes if you get too engrossed in the checker game.

Of course, now as back then, the potbelly is just one of myriad stove designs developed over the years. Stoves come in all sizes, from a tiny little foot warmer you filled with coals to keep away the chill during winter meetings to big furnaces that take four-foot-long cordwood and can power a central-heating system. Designs vary too from ornate Victorian parlor stoves to austerely simple Shaker designs. Construction materials include steel, stone, brick, and ceramic tiles as well as cast iron. You can get them in plain black iron or in fireproof enamel in any number of colors. And the variety of functions inherent in stove designs is greater today than any time since the real heyday of stoves, the nineteenth century. They come open so you can see the fire, or closed tight for most efficient combustion (and some have both features). You can find wood heaters with or without cooking tops. A few small European designs can even double as ovens and then there is the kitchen range, a separate topic altogether.

Why don't we go somewhat into the capacity of heating stoves in general, and try to come up with some guidelines to help you select a location, size and design to meet your needs. Then we'll discuss the major designs available, and how to install and operate them (safely above all).

One thing we'd best put right out front, though. There are *no* absolute truths in wood heat—except for the safety rules. Every woodchopper has own grip on the ax, every wood heater has his/her own favorite stove or fireplace, and an individual way

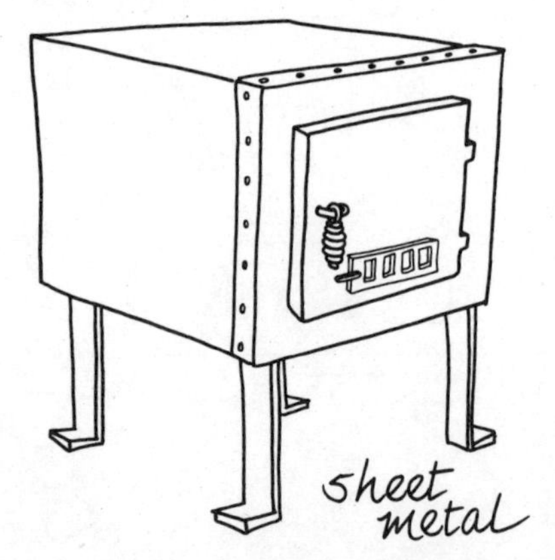

of firing and cleaning it, a single-minded preference for kind and seasoning of wood, ad infinitum. And, isn't it great? The technology is so nontechnological that we can all be our own experts, all just as right as the next ash sifter, even if we differ in our opinions by the breadth of half a woodlot. I'll try to lay out the generalities where I can. But when I express a preference for stoves such as our Jotul #4 Combi, don't you pay any attention. I can show you four dozen wood heaters who think I'm nuts; they swear by their Shenandoah or Ashley Circulators, their Lange or Styria airtight, their potbellies or their Riteway central heaters. If there is any group of people who "do their own thing," to use an overworked phrase, it's wood burners. Now, let's talk about wood stove location, hopefully, without too much argument.

Locating the Stove

The first location you might consider is to put your stove in the cellar, if you have one. Or perhaps put one of several there. One notorious disadvantage of heating with stoves and other space heaters is "cold feet"; without a central furnace or other heater in the basement, your floors are going to be icy cold in winter everywhere on the ground floor but near the stoves. The upper stories will have warm floors, though. Though we do have a cellar stove, we don't use it to warm the floor constantly, just when we want to work in a warm basement. Sheepskin slippers keep our feet warm, and the children's playroom is on the second story where they can go to sleep on the warm floor if they want without a chill.

Heating Efficiency

The best first floor stove location surely must be smack in the center of the most-used room in the house, with a register or grating cut into the ceiling above to let warmth into the second-story room where baby sleeps and grandmother beds down when she brings her easily chilled bones for a visit. With a stove in the room's center, heat will radiate out all around, then rise to the ceiling, flow outward, then as it meets the relatively chilly walls, it will fall and flow back along the floor to be warmed up again. The flow is sort of donut-shaped.

Few people will want a stove dominating the center of the living room (though that's where it should go out in the shop in

donut pattern of heat and air flow

your garage if at all possible). Next best would be near the center of the house by the stairwell going to the second story if you have one. That way, heat will move throughout the whole house, though the room in which it is located will be the warmest. Until I discovered how much plaster and lathing and woodwork I'd have had to remove for a safe installation, we planned to put our big Combi that combines the advantages of offering an open fire in the evening, but closes up for efficient operation at night, directly into the mid-house flue, even if it did serve the central furnace too. That way the whole house would have been better warmed—as the location abuts a big door leading to the stair. But, we had to settle for another location and for reasons of economy and ease of installation, we decided to put it into a new cement block flue on an outside north wall.

Now, the Combi heats air that circulates around the lower story, mingling with heat from our wood-burning kitchen range. Some gets into the hall and on upstairs, but not much. So we cut registers into the bedroom right over the stove; Sam opens it when he gets chilly, closes it when hard play warms him overmuch. There was already a register over the kitchen range; baby Martha's crib went right over it. Now that she's a big, grown up two-going-on-three-year-old, and can stay well tucked under a set of covers the night through, her bed is moved to the back wall.

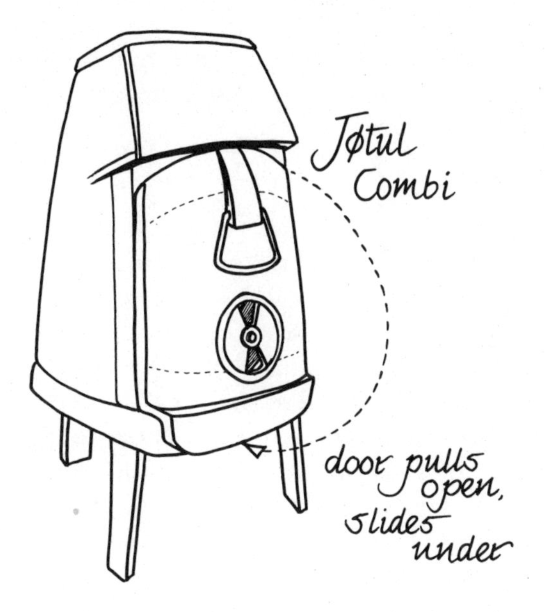

Tying Into the Flue

Stove location will also be affected by your home's chimney situation. They can go into existing flues—either in old stove holes, new ones or into a fireplace. The illustrations and photos show good examples of each. You may find that a preferred location on an inside wall justifies buying a prefabricated metal flue. On an outside wall, you can build an outside, ceramic chimney as we did. All the problems of wind direction and velocity brought up earlier will factor into your choice of flue and stove location. Just remember that a stove functions exactly the reverse of most central-heating systems, where radiators or registers are located along the outer walls of the building. There, warmth pretty much circulates up and down the wall, keeping wall and adjacent air warm. Your stove, a single, centralized, radiating heat source, pours its heat out at whatever or whomever is sitting near it, but most of the energy flows straight up, along the ceiling to a cold wall and down again. There are several two dollar words that describe these patterns of heat and air flow, but I've never been able to keep them straight. Just remember that donut pattern of air flow, and position your stove so it gets warm air where you want it most. And remember, the air at the extremes of the donut isn't necessarily cold—especially if your stove has enough heating capacity for the rooms. It's *relatively* cool, compared to the 1,000 degree (F.) plus temperatures inside the stove.

How Many Stoves?

Now, you must decide how many stoves of what heating capacity you will need to serve the space you want to heat in the position or positions you've selected. There isn't much really authoritative data available on wood stoves' heating capacity. The stoves were developed before those big words I just referred to were coined. In the old days folks picked their stove from knowledge borne of generations of experience. Today's manufacturers can and do make whatever claims they wish. One retail dealer I know of claims that the big expensive imported model he sells will heat his 11-room house without blowers to move air around or any supplementary heat. It depends on how you define heat, I guess. If his house is well-insulated and the stove is fired until it shakes, none of the pipes will freeze. But anyone trying to sleep in a distant third-floor bedroom will come pretty close without plenty of blankets.

It isn't as well known as it should be, but the heating capacity of a stove is a direct function of the stove's size—both the amount of surface area it has to radiate heat and the pounds of material in it. You'll notice that older American cast-iron stoves are all numbered; the number refers to the weight, the pounds of cast iron in them. Till the recent oil crisis regenerated demand for stoves, and prices went skyrocketing, they were sold by weight too. A dollar a pound at the beginning of this decade; this was up from about a dime a pound back in the 1880s. These days, of course, you can't touch a good-quality, new cast-iron stove for less than a few hundred dollars, no matter what its weight.

Now, how to figure the size and number of stoves to keep you warm. Sorry to get mildly technical, but heat is measured in British thermal units, or Btu's. One Btu doesn't amount to much: it's the amount of heat needed to raise one pound (pint) of water one degree Fahrenheit. Heating capacity of furnaces is calculated in the number of Btu's they can put out per hour. If you already have central heating and would like a rough idea of how many Btu's you are using, check the specification plate on the front of your furnace. Gas heaters give the Btu's put out per hour. Oil furnaces list an oil consumption rate. Multiply this by Btu's in a gallon of #2 heating oil, 140,000, to get the overall heating capacity of your present system. If you've an all-electric heating system, I'm sure you are all too familiar with your Btu needs as you've tried to keep those astronomical bills down to something approaching a reasonable level.

Figuring Heating Needs

The furnace that came with our house has an oil-burning rate of 1.3 gallons an hour to heat both the house and the hot water supply. At that rate we would use, without the wood stoves, a bit less than 1,000 gallons a season. We figure the furnace would be burning at an average of a quarter of the time, year-round, but during extremely cold periods it would be going at full capacity. About a quarter of the energy goes into domestic hot water (based on the approximate relationship between hot water and space-heating costs in homes with families of our size that have separate water and heating systems). This would leave a gallon per heating hour or so to heat the space when temperatures were at their coldest, or 140,000 Btu's per hour. For our seven-room house that comes out to a convenient 20,000 Btu's maxim heat required per hour per room. From what little information we can find on space heaters, I gather that this 20,000 Btu per room figure is fairly accurate and typical. Not to say that you would be needing all that heat all the time, but to stay comfortable in all kinds of weather you should have about that much heating capability. And the figure applies to rooms up to twice the size of our parlor, 15-by-18 feet or 270 square feet on the floor with an eight-foot ceiling, or about 2,200 cubic feet. You might say, you need five to ten Btu's per cubic foot of air to be heated depending on how warm you want it.

Stove Size and Heating Capacity

In fiddling around with weights, surface area, general construction and heating capacity of a few dozen wood stoves, we've reached the following conclusions. For all types of iron, stone, and other massive stoves, heating capacity is a diminishing function of the stove's weight. A little stove, weighing in at 125 pounds, will easily heat one room of our parlor's size, but if you push it, or are very well-insulated, it will heat twice that amount of space. Double it in size, to 250 pounds, and it will only heat one more room, triple it, and you only get another room.

From our experience, a little iron 100-pound stove, if of an efficient design, will hold a fire overnight, but can't hold enough wood to keep anything more than a small room lukewarm. Also, simply because of its smaller mass, it takes more wood than a big stove to generate a large amount of heat. On the other hand, because of the diminishing return on increased size, a 300-pound stove will not be three times as efficient a fuel user or heater as the little models. So, in our home we could have gotten about the same amount of heat from two 100-pound stoves as from the single 300-pound Combi. Plus another set of stovepipes to clean and ashes to clean up. The single big one is best for our square, relatively large-roomed house. Folks with other layouts might do best with a pair of 100-pounders, added chores included. Another advantage of the bigger stove though, is its ability to hold a lot of wood. The bigger the stove, the more fuel it will hold, thus the more heat it can turn out during the night. For example, on below-zero nights we can put in five or more six-inch diameter maple logs, close the damper down and adjust the draft so that each 10-pound log lasts about two hours. We'll get up to a good house-warming 30,000 Btu's the night through, even if we all sleep in in the morning. A smaller stove, equally as efficient, would take half the amount of wood and need to use it almost twice as fast to turn out that much heat. On the other hand, both the big and smaller models will produce half that amount of heat (about 15,000 Btu's) on half the amount of wood (some two-and-a-half pounds an hour), a fuel consumption rate that will keep the chill off a small house in not-too-cold weather all night long.

With the less efficient potbellies, and Franklin-style stoves where air input can not be strictly controlled, heating capacity is also a diminishing function of weight. Our little 50-pound kitchen stove back on the farm was fine for right there, but it didn't project heat into the drafty west end of the room. A larger stove, such as the Franklin in the homestead living room—in the 100 to 200-pound range—did a good job of heating a fair-sized room. But even a 300-plus-pound conventional stove won't heat more than two rooms. None will hold a good heating fire overnight.

In welded steel and sheet metal stoves, you aren't so much interested in weight, but in surface area. Obviously the bigger the stove, the larger the fire and bed of coals it can hold and the more surface there is to radiate heat. One of the little sheet metal jobs you can get for under $50, even today, will put out your room's worth of heat, 20,000 Btu's if you need it, but you have to keep a good fire going constantly. My calculations suggest that a steel stove should heat about half as many cubic feet of air as there are square inches of radiating surface on the stove. This is the same figure that applies to the largest cast-iron stoves, by the way. The smaller and more efficient sheet metal and welded steel models will heat perhaps half again as much, the number of cubic feet equivalent to 75 or 80 percent of the number of inches of radiating area. I guess what all this boils down to is that 100 pounds of cast iron will heat two average-sized rooms if it's efficient, or only one if it's not. Each added 200 pounds is worth two rooms in an efficient stove, but only one in a conventional one. And in steel stoves, the bigger they are, the more heat they'll throw. I don't know why I spent two days cussing over a slide rule and calculator to figure that out, but maybe it will help someone.

Heat Savers

While we're at it, now that we are into Btu's, we should mention the amount of heat you can rescue from the stovepipe with one of those $100-plus heat-saver gadgets you must have seen advertised. They run flue gasses around boxed, open tubes, then blow room air through the tubes. We'll discuss them in more detail later. The amount of heat they can pull depends on the temperature of air in the pipe. With a good hot fire, it can approach 1,000 degrees F., and at least one manufacturer claims its biggest unit can reclaim over 50,000 Btu's per hour at that temperature. I'll stick with the more conservative claims I've

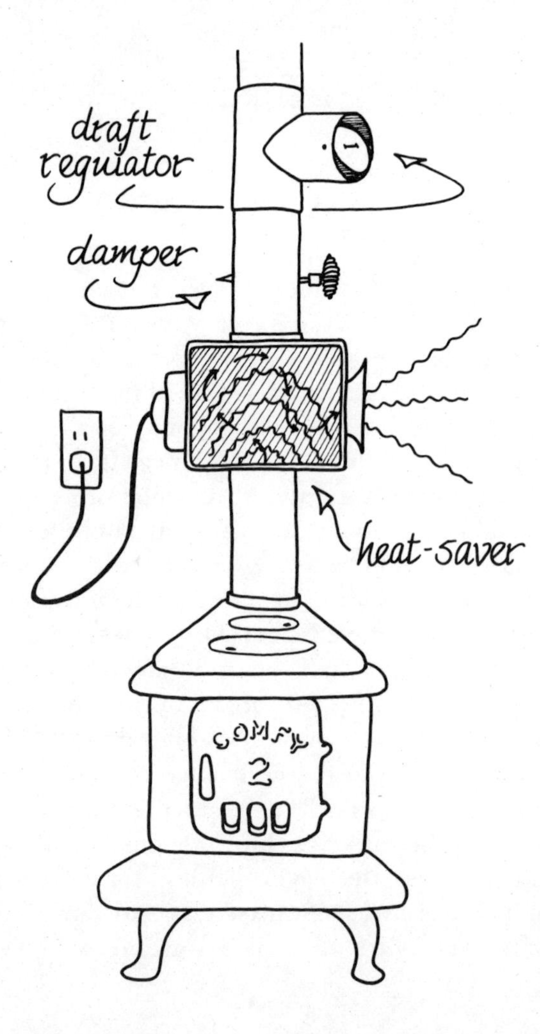

seen—from 10,000 to 20,000 Btu's per hour depending on temperature of the stack gasses. Of course, the less efficient your stove, the more benefit you would get from the heat-saver. I hardly think they would be worth the money if yours is a Scandinavian, circulator, or other airtight design. But with a big, roaring Franklin you may actually get more heat from the saver than from the stove. If we assume for convenience's sake that a saver saves an average of 14,000 Btu's per hour, in a 24-hour cold snap, it would retrieve 336,000 Btu's. If a gallon of oil is worth 140,000 Btu's, that comes to about two-and-a-half gallons of oil (selling just now for 40¢ per gallon) or a good dollar a day. The unit should more than pay for itself in a single heating season.

Circulating Stoves

Now, what about the circulating wood heaters, the kind that come in contemporary-styled cabinets and have thermostats, patented downdrafts and all? They are as capable as the efficient conventional radiating kinds. One advantage is the (usually) optional fan that blows air out and circulates it into the house. With the blower, most manufacturers claim their units will heat an average-sized house and I know plenty of New England homes that have been heated with circulators for years. However, I admit to an utterly unjustified bias and wouldn't have one in the house. Reason? The fuel doors are on top or at one side and you can't see the fire, which you can by opening the front-located door of any radiating stove. You can't cook on them since the cabinet never gets warm enough to even heat up winter-chilled feet. You never even smell smoke, they are so carefully engineered. It's same as having an oil furnace in the room. Only difference is you have to load it with logs once in a while, and empty the ash tray.

I must admit, a lot of folks would find these features positive advantages. That plus the fact that the heating contractor who sells you the stove has charts that will tell him which model is best for your needs. Just as with conventional central heating, room size, type and amount of insulation and siding, whether or not you have storm sash and doors all enter into the equation. So, get the big model circulator for a big house, a little one for a little house, set the temperature you want and keep adding wood. It's the only way to have wood heat without a lot of experimenting

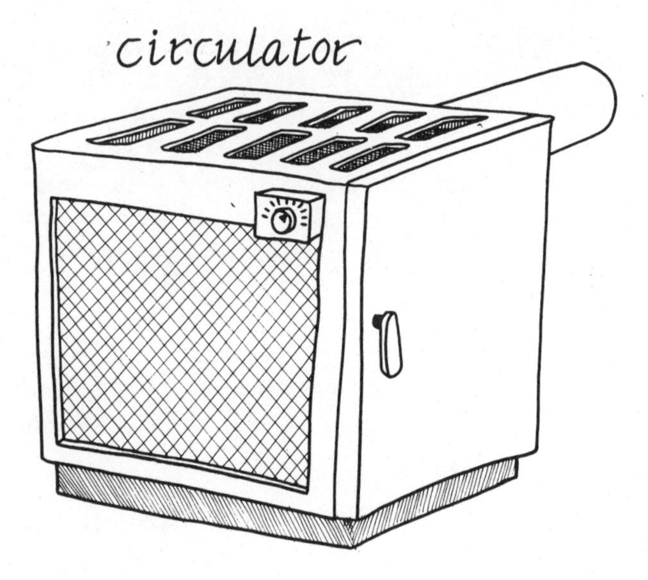

and not a little shivering and sweating. But no fun either, the way I see it.

While we're on the circulators we may as well consider their operating principles, the plusses and minuses not mentioned above, then do the same for other stove types you may want to consider.

One great advantage of the circulators is the near-automatic operation. They are probably the safest designs of all too; a spark simply can't get out of one, and if you keep the automatic damper mechanism in good condition they can never overheat. As stated above, the outsides do not get hot enough to burn, which is the case with any radiating stove. If there are babies in the house, particularly if you've more than one crawler or toddler and can't keep an eye on all at once, a circulator can save you a lot of worry and potential grief. The metal of a well-fired radiating stove can take the skin right off a little hand, and don't forget it. Kids do get wary of the stoves in time, and at a surprisingly young age. But until they do, a newly installed stove is a hazard—*except* for the cabinet-enclosed circulators.

Most circulators have firebrick linings, cast-iron grates and sheet metal outer cabinets and will last forever. Though each brand has its own method of introducing air for combustion to

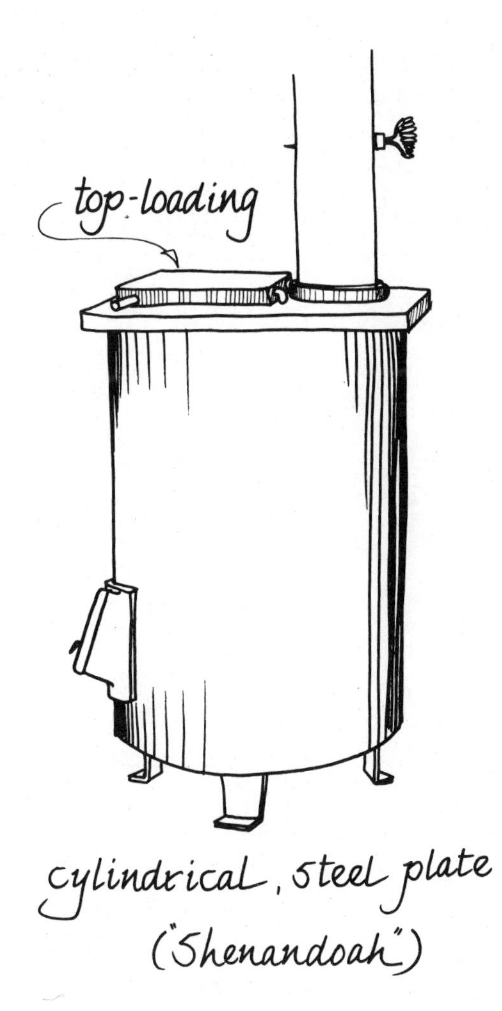

cylindrical, steel plate ("Shenandoah")

the fire and blowing room air around the hot interior and into the room, some of which are patented, the principles are pretty similar. (One of the best and widely known brands, Ashley, has a patented "downdraft" system for warming combustion air. I don't suggest that they are attempting to mislead anyone for a second, but this is not the kind of downdraft Ben Franklin tried in his coal stove — and that is most successfully employed in central heating furnaces to force gasses produced by the fire back into the coals or flame so fuel will burn as completely and economically as possible.) Here is a generalized description of the way circulators operate.

The firebox usually has fire door and ash removal from one

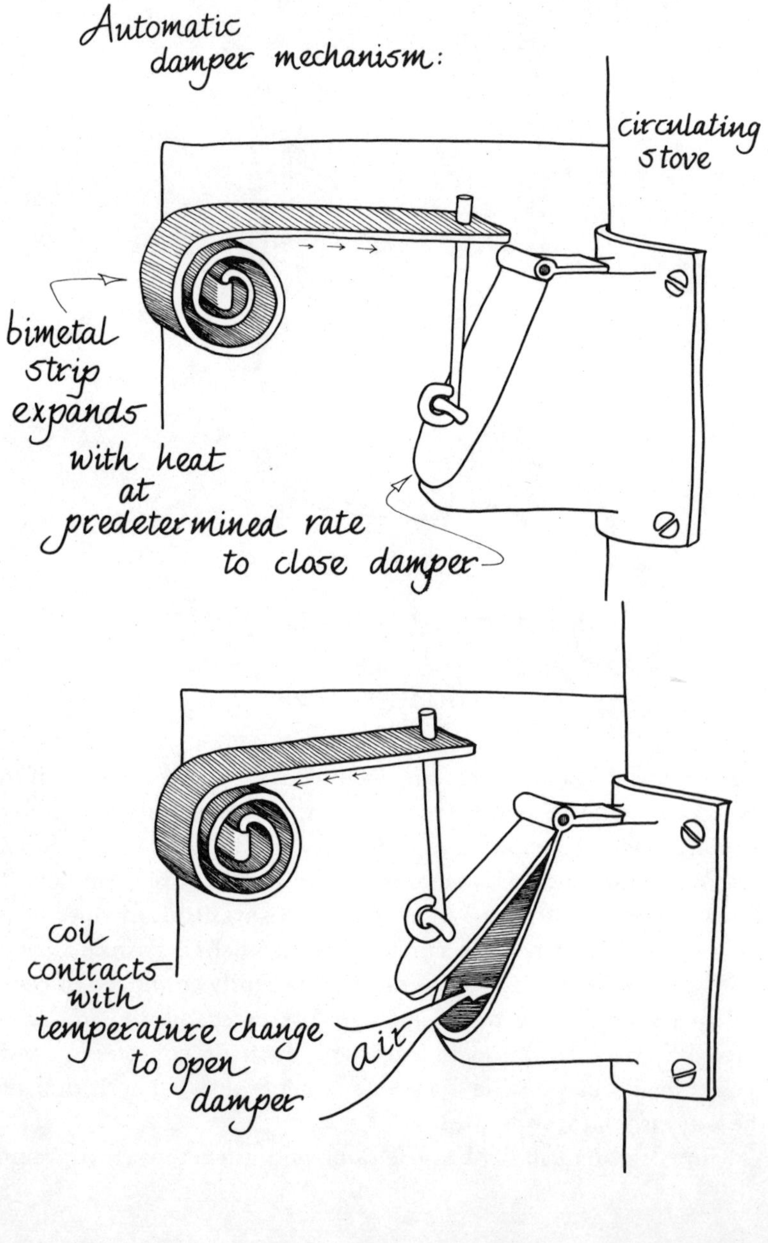

side, smoke hole in back and thermostat in front. Air for combustion normally enters at two levels. Air for primary combustion at the level of coals enters at the bottom, while air for secondary combustion of gasses enters at the top. The amount of air entering is regulated automatically by a bimetal strip, two different metals fused together in a coil or other spring shape. As

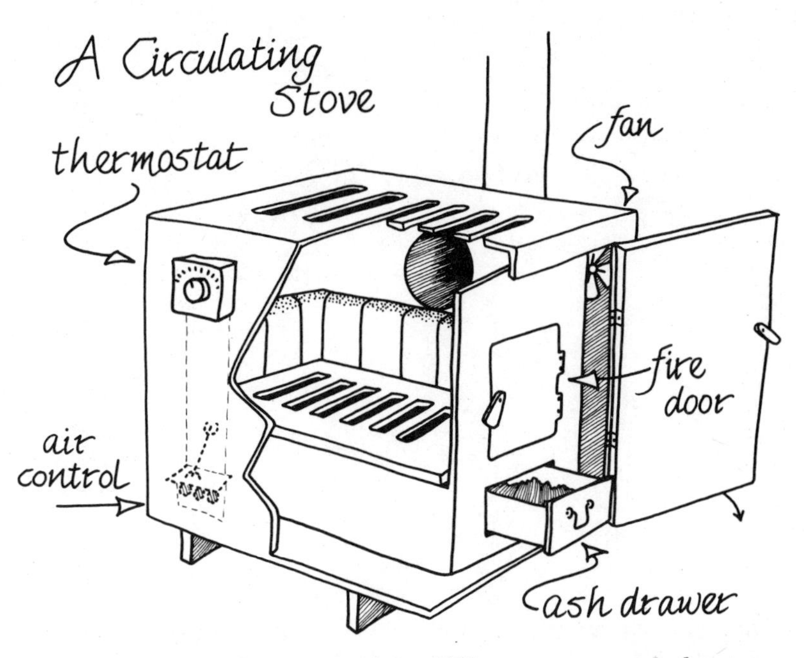

they are heated, they expand at different rates and the stress causes the strip to expand at a predetermined and constant rate. If you set the thermostat to have room air at 60 degrees F., the strip will bend back and forth, closing or opening the draft control to maintain a fire that will keep the heated air flowing past it just at 60 degrees F.

Room air is forced around the hot inner firebox by gravity or a fan attachment, all very automatic. Some brands have complicated smoke aprons and ash screens that close when the cabinet and firebox are opened so you never smell smoke or see a flame or get ashes on the light beige wall-to-wall carpet. As I say, I wouldn't have one in the house. Louise wouldn't have light beige wall-to-wall carpet either. But if you do, only a circulator will let you keep it clean without a lot of extra precautions.

Scandinavian Airtights

The bimetal-strip type thermostat is found on a growing number of conventionally designed domestic and imported stoves, both cast-iron and welded steel plate models. All are carefully made with sealed joints so they are properly termed "airtights." Only the amount of air needed to attain the temperature you select is permitted into the fire—all automatically.

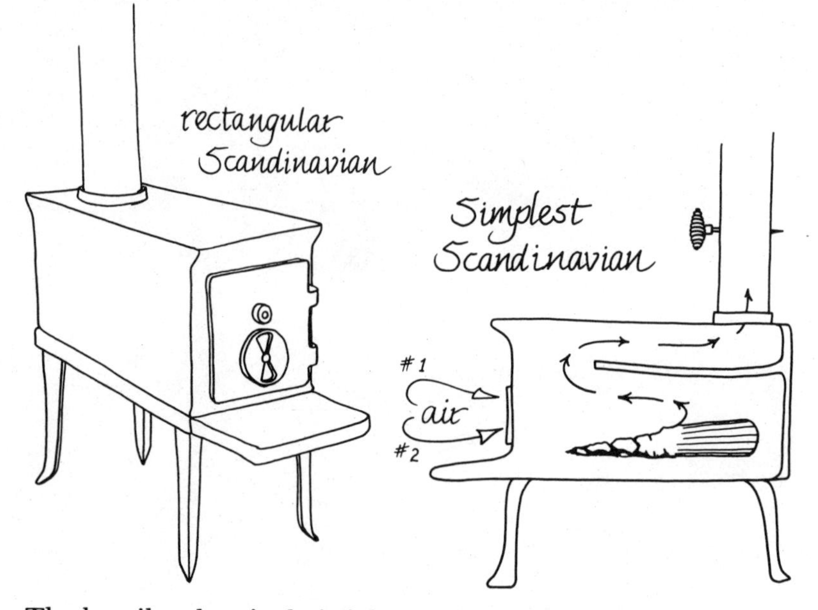

The heavily advertised airtight stoves made in the Scandinavian countries—Jotul, Morso, Trolla—and some models from Central Europe neither have the bimetal thermostats nor need them. The principle they use is based on complete airtightness and one or another system of interior baffles that force the smoke into a serpentine pattern of movement inside the stove. The simplest is a single plate extending from the back of the firebox to about three-quarters of the way to the front. In these stoves wood is advertised to "burn like a cigarette" from front of the stove to the rear, and it does. The only air that is admitted comes in at two levels from the finely machined air draft control for primary (1) and secondary (2) combustion.

All Scandinavian airtights are beautifully cast and assembled. The door and opening faces are either finely ground or fitted with asbestos rope gaskets and all have latches that close them securely. They come in standard black iron that needs occasional "polishing" with stove black, in black fireproof enamel and in an assortment of other enamel colors that may look strange to a non-Scandinavian eye. Jotul offers a deep forest green and other brands come in unfamiliar (to us) shades of rich red, blue and other colors. The enamel is pretty tough but can chip. We prefer flat black in an iron stove because a coat of stove polish can repair any discoloration. The plain ones are cheaper, too.

If there is a disadvantage to the Scandinavians it is cost—they are expensive. And, since they are manufactured in countries using the metric system any replacement nuts for provided stove bolts must be metric sizes. So must the stovepipe, or at least the section fitting to the metric-sized smoke boot. Metric pipe is also imported and is expensive. Some manufacturers sell metric pipe in special heavy-duty gauge steel. At least one U.S. manufacturer is putting heavy-gauge pipe together in U.S. sizes. It is expensive too, but worth it if you would find replacing your stovepipe every few years a nuisance. If you can't find it where you buy your stove, look through the mail-order catalogs offering stoves. You can purchase adapters, or improvise as we did. In adapting metric to inch pipe or any other change in pipe size always go up in size—get the nearest larger size, never smaller.

Unlike most U.S., Canadian and Central European designs, the typical Scandinavian stove lacks a grate or log rests. The manufacturers recommend building up a deep ash bed to protect the bottom from burning out, but I'd suggest putting at least a

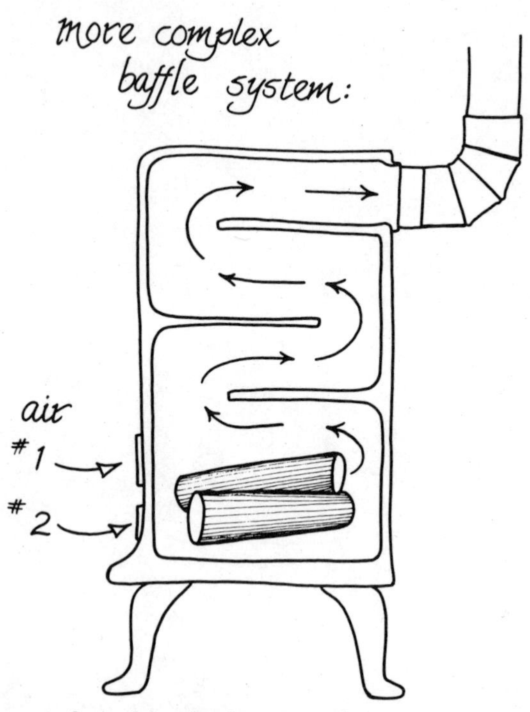

two-inch layer of sand in the bottom of any grateless stove. Particularly if it's an expensive import.

Creosote

Before we leave the airtights, imports, domestic or homemade, we'd best mention their greatest drawback, and this

applies to circulators too—in spades! When closed down, the fire is flameless. However, because of the very limited amount of oxygen entering the firebox, combustion is incomplete. Several components of the wood vaporize without burning in the intense heat, and escape into the flue. Upon reaching the relatively cold stovepipe or flue liner they condense into a fluid called pyroligneous acid, which Ben Franklin wanted to burn in his unsuccessful downdraft stove. This is a brown sticky substance consisting primarily of methanol (wood alcohol) and acetic acid, the same stuff that gives vinegar its bite. It can drip from your stovepipe and stain the carpet, but is otherwise harmless in itself. However, the acid tends to dry and bake on the insides of pipes and flues, becoming what old-timers call creosote. If you're interested, the acid is one component of the commercial creosote you buy for such things as preserving wood. Commercial creosote is obtained by distilling wood, beech in particular, by heating it in the absence of flame to the point that the several aromatic oils in the wood turn to vapor; they are pulled off and cooled, whereupon they condense into creosote. This is precisely what happens to part of the wood in the flameless firebox of your airtight stove: you are distilling the logs at the top, the ones up and away from the burning coals.

The Danger

Once dried out, creosote is flammable stuff, though it takes a lot of heat to set it afire. I've seen plenty of antique flues with many inches of creosote buildup on them. The relatively cool, inefficient fires of old damperless fireplaces had a comparatively hard time getting up to a creosote-tindering temperature. But plenty of them have done so; the dozens of cellar holes scattered through our New England woods attest to that. However, the danger of a flue fire is minimized if you have a modern chimney with sound brickwork and liner. Chances are the flue fire could burn itself out without destroying the house, though the great fountain of flames showering the neighborhood with sparks will bring the fire department in a hurry. In an older flue with cracks that might let fire out and into the house timbers, a flue fire can be a disaster. In any event you should avoid them if possible and have that fire extinguisher mentioned earlier good and handy to control one, if and when.

Beating Creosote Buildup

All airtight stoves will produce creosote if operated at maximum efficiency in the closed-down mode. Frankly, it scares me to see so many friends putting them into the old flues of antique homes; if I have any say in it, they get an experienced mason to go over the chimney from ground to roof before they light up. And, no one should attach an airtight to a extended length of stovepipe. One of our photographs shows the loose creosote accumulated in the pipe of an old turn-of-the-century airtight in our house. Had that creosote ever caught, the tremendous demand for draft would have sent the pipe into gyrations; it would have torn itself apart, hurling chunks of flaming creosote all over. As we've said before, long pipes are OK for some installations, but not for airtights. Plan to vent yours directly into a flue.

Now, considerable creosote buildup can be avoided by having an open fire during part of each day. This is one reason we chose the Combi for our main wood heater. It is kept open during the day, an open fire going most of the time, with the damper nearly closed. Then we pile on the wood for a high-flame cheery fire during the evening. The stove is closed up for most efficient, (but creosote-producing) heat only during the night.

There are several products on the market that purport to remove creosote from your flue. They go by such names as Fluesweeper. First of all, they are nothing but rock salt that you can buy by the 100-pound sack for loose change. And they work by vaporizing the salt which combines with the creosote to make it catch fire. Perhaps if you threw a handful of salt on the fire each day and burned the creosote as it accumulated, it would be safe. However, no government agency or fire prevention association that we contacted recommends using the stuff, so neither will we. If you are going to install an airtight stove, do your planning as if a flue fire is inevitable, then do your best to avoid having one by cleaning the flue at least once a year. And, during the day, have an open fire when you can—door full open.

Flue—Fire Alarm

While we're on the topic, you'll find no commercial flue fire alarm on the market. But a friend happened upon this dandy alarm in an old book of farming information:

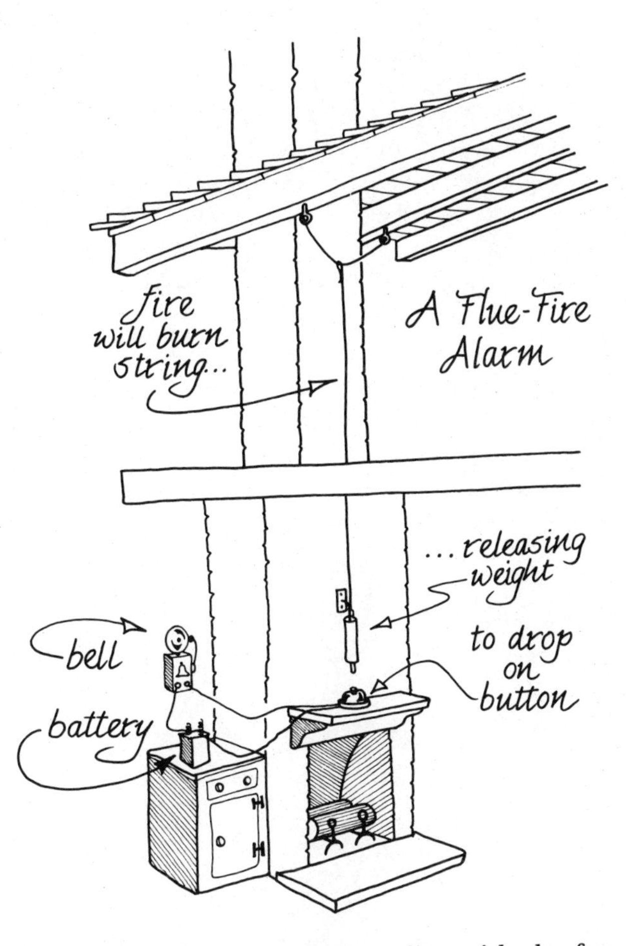

"Drive a nail in two rafters on a line with the face of the chimney, to which stretch a cord close to the chimney, so that, in case of fire, the cord will burn off and release the weight hanging to it, which in turn will drop on an electric button and ring a bell. A dry battery will cost 20 cents and a bell 50 cents. Place these on a shelf above the fireplace. Place a piece of heavy wire, 10 inches long, as shown, and fasten to the wall or chimney for the weight to slide on. The weight need be suspended only an inch or two above the bell."

There really isn't a great deal to say about choosing a conventional nonairtight stove that hasn't been said about airtights. Be sure to put a few inches of sand in the bottom of them all, sheet metal ones in particular. Then, if you are a serious wood heater, trade it for an airtight.

Buying an Old Stove

Probably the best advice to give on the topic headlined above is "Don't." For one thing you will probably have to pay a premium for the age factor, and just about any old stove will be used—really used. In the old days they fired stoves till they literally fell apart. Most spent at least the latter part of their first go-round burning coal, which generates a heat intense enough to burn out even cast iron in time. Further, no single book on earth can list the names of the huge variety of stoves that have been manufactured over the years, to say nothing of telling you how to run them or what parts they need (and which may be missing). However, everyone would rather have one of the old-timers if he can find a good one (we have six at present count) so, here's what I'd look for in the used stove shops.

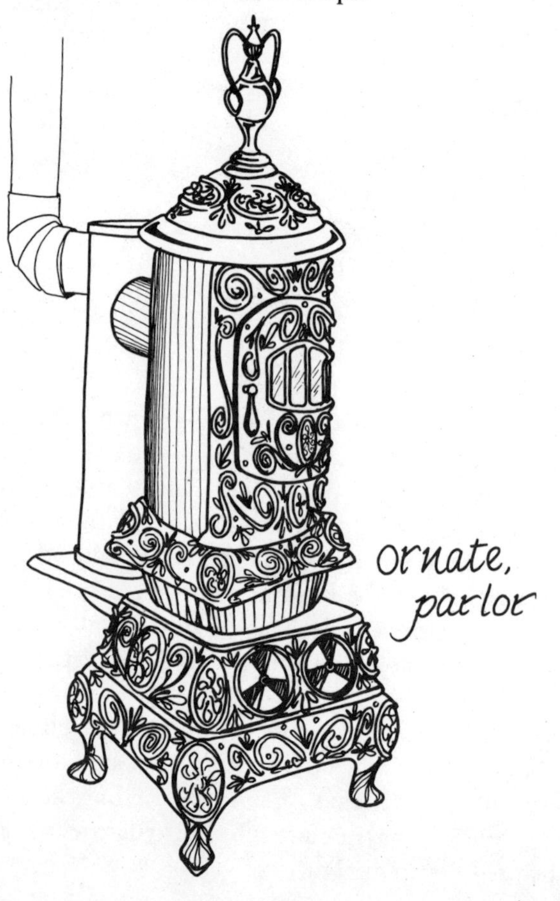

Our house came equipped with a couple of old wood-burning stoves, one of which was this 1898 Cole's Airtight. By today's standards, installation wasn't safe, as the stovepipe entered the flue too close to the ceiling and was too loosely wired. But it heated a part of the house for more than 75 years, run by a fellow who learned wood heating a century ago and knew how to keep the fire in bounds. We decided to rejuvenate the stove and install it in our basement workshop.
The first step was to take down the stovepipe, which had several years' accumulation of soot in it, and move the stove to the outdoors for disassembly. Many rusted soft-iron stove bolts had to be cut with a chisel, hammering gently to avoid splitting the

old iron. The soft iron liner was burned through and came right out. The brightwork wasn't at all bright, so it was sent to a chrome/nickel electroplating shop—the sort that trades with antique auto buffs—for replating. After the stove was lined in galvanized steel at a sheet-metal shop, I lined the firebox with soft but heat-impervious K-26 firebrick. The bricks I carefully trimmed to fit, then cemented in place with refractory cement. The top and the brightwork went back on, the bottom

was covered with several inches of sand, and the stove was ready to hook to a flue. It all sounds easy, but it took a lot of cutting, drilling, punching, fitting and cussing to get everything together airtight. With the stove at the ready, I next chiselled a hole in the flue in the basement and cemented a thimble in place for our "new" basement heater. The best approach here is to chip out most of the mortar surrounding one center-course brick. Then split the brick and remove it. Finally, chisel away at the adjoining bricks until you have the right-sized hole for your thimble. The thimble is an adjustable metal sleeve that goes in the flue, making installation and removal of

the stovepipe easy. And since the Cole's Airtight is positioned in front of the chimney cleanout, ease of stovepipe removal is significant. It was a whole lot of work, but with the old girl newly blacked, polished, installed and fired up, it almost seems worth it. I'd bet she'll last another 75 years.

Bonding Breaks, Patching Holes

I can't think of much about buying old sheet metal stoves that isn't covered in the repair job on the Coles airtight. Look for rust inside and out. Be ready to replace the liner if there is one, and to patch holes if there are any. You will have to use nickel welding rods and patching metal capable of withstanding at least 1,000 degree F. temperatures. A welding shop has the equipment and know-how to handle it, and since I don't, I'd take any patching jobs to the pros. If you are one yourself, fine. But patches put on with the little propane torches (or even the new hotter-firing welding gasses such as BernzOmatic's Mapp Gas, which I use around the house) could come unstuck on a hot stove. Nickel welds will hold, but I would never try patching a sheet metal stove myself. I wouldn't trust the mend to hold up under all eventualities, such as overheating, and I wouldn't feel confident with the stove. Put patches on with equipment capable of cutting through boiler plate. And, frankly, sheet metal stoves were never very costly or well-made things, even in the old days. Unless you find a genuine rarity, I'd say that extensive patching of an old sheet metal stove isn't worth the time and effort. Besides, if it is old enough to have an antiquarian value, the metal just may be too corroded or soft or flimsy to stand up to the heat needed to apply a safe patch.

Cast-iron stoves are something else. Some of the old beauties are worth a lot of work to get them back in shape. But one warning: you can not rebond broken cast iron so it will hold a fire. I don't care what anyone tells you. Sure, there are modern space-age glues and welding techniques that will mend a break so a stove leg, say, will hold weight and look fixed, but it isn't. You see, as it heats up, cast iron expands at a rate that it feels comfortable with. Nothing else expands at quite the same rate, and if you try to put a layer of glue or welding bond between one piece of cast iron and another, the three pieces will expand at three different rates and you have a break all over again. The only way to repair cast iron to really last is to replace the entire casting. Some stoves are common enough that you may be able to find two or three with good salvageable parts to go together in one good one. Or, for a price you can have a foundry cast you a new part. You'll probably have to look and talk hard. Few foundries are in the small-job businesses. Many no longer use the split-box, sand casting technique needed for most custom work. But do look and talk. And be prepared to shell out. Take the broken part to the foundry boss for instructions on how to make a pattern for their mold. You may find that they will prefer to break up a large casting into several parts. You'll have to work up the patterns in wood, clay or plaster of paris. It's a lot of time and effort, but worth it. Usually.

Sometimes you can rig bolted-on arrangements to hold broken cast-iron pieces together, but, frankly, if an old iron stove is valuable enough to deserve repair, I would want to do my best to recast any broken parts that I couldn't repair. But before going to the trouble of making up foundry patterns, I'd check a lot of antique stores for a similar stove. Also, some of the modern stove works have inventories of old parts. If the maker of your stove is still in existence, I'd give him a try. (Just don't bother the Glenwood people if, like us, you have one of their old stoves; they don't make wood stoves any more and have no parts, or so they told us when we asked. Actually, maybe we should nag them and other former wood-burner makers. Maybe they'd go back into production.)

Uncovering the Flaws

Many stoves will be missing parts, and just what to look for we can only discuss in generalities. Of course, you want all doors, lids and such to be there. Don't worry about inside grates that may be burned out in a stove that was used for coal. For wood use you can support logs on brick or adapt a log rest, andiron or the like to the stove. Do be sure the body is sound, though, particularly the floor and sides of the firebox, which are the most likely parts of the body to be burned out or cracked. If the stove has grates holding the fire well above the bottom, a rusted-out floor or cracked side can be fitted with a sheet metal liner; add the sand layer to the bottom and the liner will serve to catch your ashes. Indeed, you could line an entire iron stove with sheet metal, so long as joints are sparkproof, but since most of the old models were cast in intricate bumps and curves, it would be quite a job. And would need replacing in a couple of years.

Holes in the body are a bad sign, even if the stove has a shiny new coat of stove black. In fact, they are an even worse sign if it

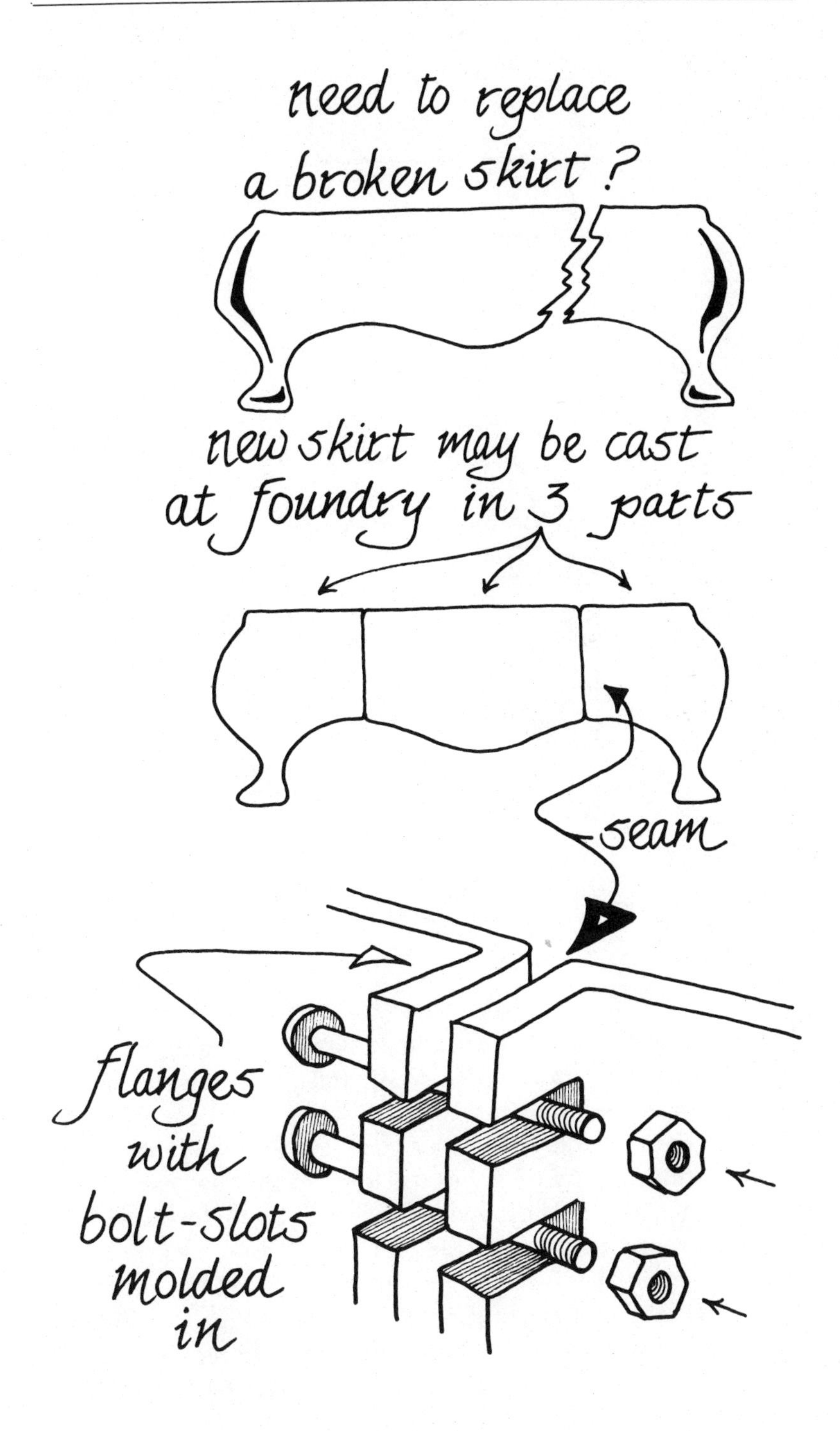
need to replace
a broken skirt ?
new skirt may be cast
at foundry in 3 parts
seam
flanges
with
bolt-slots
molded
in

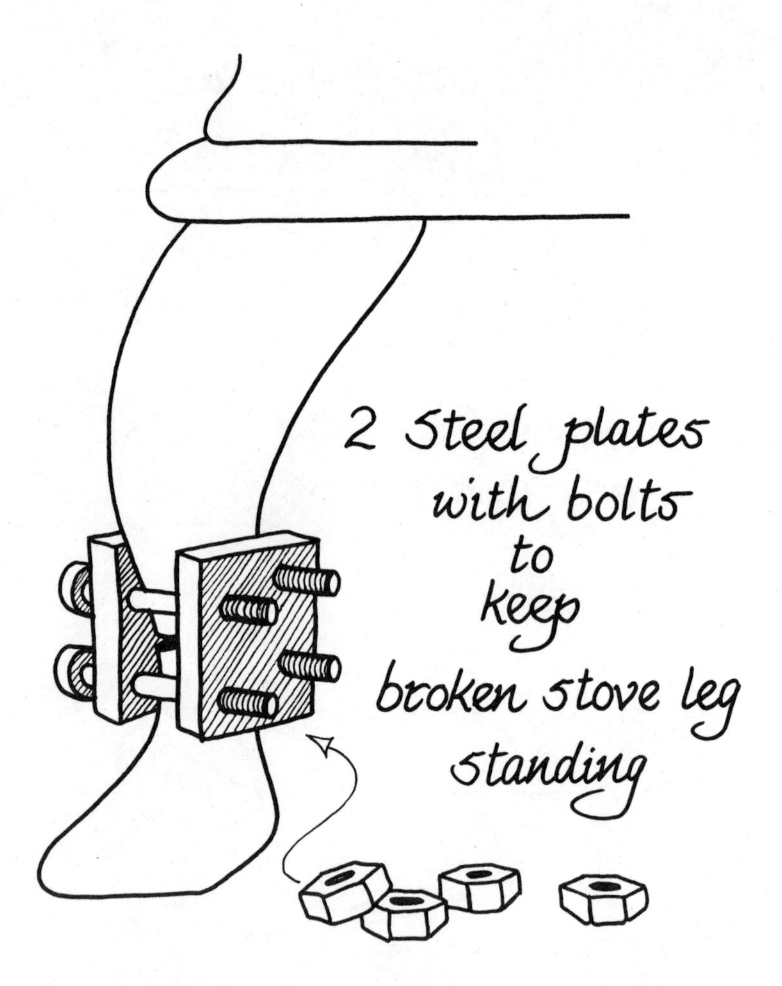

has been prettied up; the seller may be camouflaging a broken-down stove. Holes can be puttied up and painted over, masking them till your second or third fire. Were I you, I'd feel free to go over the entire thing with a hammer. But go gently; there's no use knocking holes in weak but still serviceable iron, particularly if the damage you do buys you a bum stove. Listen for an unusual thunking sound rather than the dull ring of iron. You may have found a patch. If it comes loose and is bigger around than your little fingernail, I'd forget the stove. You can keep plugging up holes with stove cement or a blacksmith can hammer in a button of wrought iron if the casting is otherwise sound, but all will work loose in time.

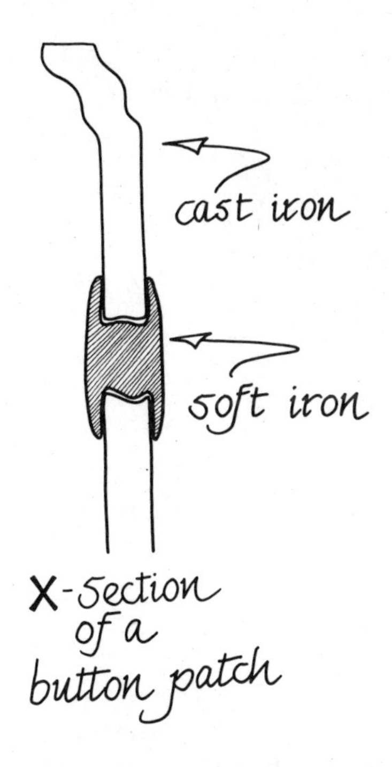

Missing Parts

Look especially well at the hinges of all doors or opening tops. A broken hinge may mean a ruined stove. Nearly all iron stoves were made up of several distinct castings that are held together by stove bolts. The bolts are of soft iron and can be chiseled off or drilled out easily. If they are rusty or missing, don't worry if you are handy with tools. Do be sure that all bolt holes are intact, or at least capable of accepting a bolt. Many stoves are held together by lengths of rod, threaded at each end; the ends go through holes cast into the two parts and nuts are tightened down to hold the stove together. These too can be replaced.

Such removable items as handles, isinglass peep windows, bright metal trim or sliding draft controls are often missing, and many can be replaced. Just be sure the bolt holes are sound. Some will be threaded—threads molded into the iron. Cast iron is devilishly hard stuff and you'll never be able to rethread a hole in it. I suppose there are diamond bits that will bore into it, but I don't have any. Unless you do, I'd not buy an old stove with the idea of boring holes in the iron to repair or replace a missing or broken part.

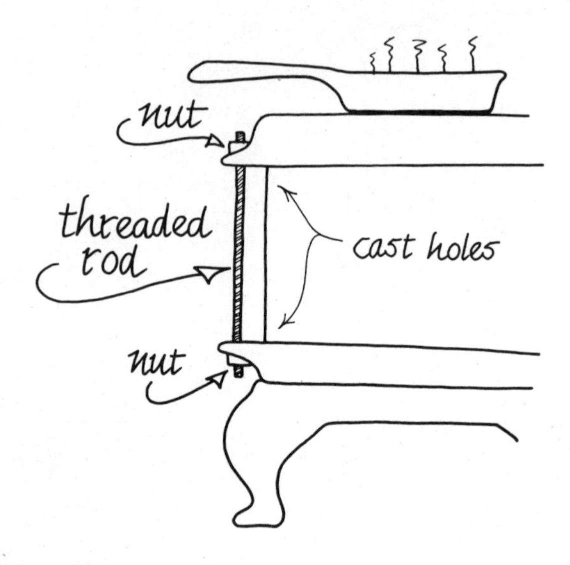

Living With the Damage

Quite often, if you are after functional rather than aesthetic characteristics, you can have replacement parts made up of wrought iron or sheet steel. This can be drilled. I know one or two blacksmiths who can take a slab of soft iron and hammer out a presentable facsimile of a missing stove leg or whatever. A welding shop could do the same from angle iron. I've seen a one-legged stove functioning perfectly well resting on three piers of brick and the good leg. A little laundry stove that we almost bought some time back had the door made of old pink Russian stove iron – a good job and if anything making the stove more interesting than if the original cast-iron door was still on.

You may find old stoves with warped plates and open seams between castings. (I'm not referring to cracks *in* a plate.) These should lower the price a good deal unless the stove is a real oldie. But, warping comes from overfiring, usually with coal, and a moderate skew to the fire-side of a kitchen range or the back of an old heating stove you'll be using with wood is not always a reason for serious concern. I think I'd plan to install a firebrick liner or find cast-iron liners if you can, and treat the old stove gently. Fire up slowly and don't ever try to get it to radiate much more than a gentle warmth. Cracks between seams can be filled with furnace cement if they are thin – an eighth of an inch at the most.

I've never bought a used soapstone stove, but I can guarantee that any real oldies you find will have cracks in the stone. Don't worry so long as the stove appears sound and the metal frame is holding together. Soapstone just naturally cracks in time. Also, don't worry about eaten-out firebrick in an old stove. Firebrick was for coal-burning models and you don't need it with wood. If it has burnt through to the metal, though, plan to put in a patch. File grooves in the old brick and apply fireclay in successive layers as thick as you can make them, till there is a good inch or two between fire and the stove's shell. Don't knock it with a stick till it's fired up good and hard. It will be powder at first. But heat turns it to rock.

So, in sum, look over any old stove carefully. Your common sense will tell you what should be there; just make sure it all is. Go through the mental exercise of firing it up, removing ash, cooking on it if there's a stove top; mentally move it into your place, attach a stovepipe and all. If something that should be there isn't, the exercise should turn it up. (Evaluating old kitchen ranges is a special problem, by the way, covered in the cooking section.) There's nothing wrong with a fresh coat of stove black, but be particularly wary of newly *painted* stoves; go over them for plugs and patches. And lots of luck.

Installing a Stove

Well, by now you've decided just what size and style of stove you want, and likely have it sitting in the center of your living room. Before installing it, consider first the cardinal rules of safe stove installation, which I must admit are honored in the breach more often than they should be—by many folks, including us before we really boned up on wood heat. I strongly advise you to follow them to the letter, even if they are more stringent than the rules in your local zoning codes.

The Cardinal Safety Rule

In essence the rules boil down to one sentence: no wood stove, stovepipe or flue should heat combustible material of any kind so hot that you can't rest the palm of your hand on it indefinitely. Now, this runs counter to a lot of old-time "wisdom" you may pick up. Such as the pre-turn-of-the century practice of using the flue to support main floor timbers in the house. Well, even a sound chimney can heat up with a too-hot fire and can bring a dry, old floor beam to the point of combustion. Today, a minimum of a two-inch gap or firebreak is built in between wood house parts and flues. If your old chimney doesn't match up, line it with pipe, or keep your fires low and stack gasses cool.

Lots of old stoves sit on high legs, and in the old days they were placed right on the floor, or on a stove board of thin wood covered by a sheet of metal. Now, one of the superinsulated circulating heaters or a cookstove where the firebox is located a good yard above the floor are safe on floors or a stove board. No other stove is. Check the photos of the several stoves in the book. All but the cookstoves rest on some sort of stone or ceramic base, and these plus legs give enough height to satisfy the quantified version of the main rule:

There must be at least 18 inches of space between a radiating surface and any combustible material.

Period! That applies to stovepipe as well as the stove.

Increase that distance to a good yard between any combustible surface and the top, back and sides of any radiating stove. The insulated circulators can go anywhere and, of course, you can back a conventional stove right up to a bricked-up fireplace with a good ceramic hearth—so long as the mantle,

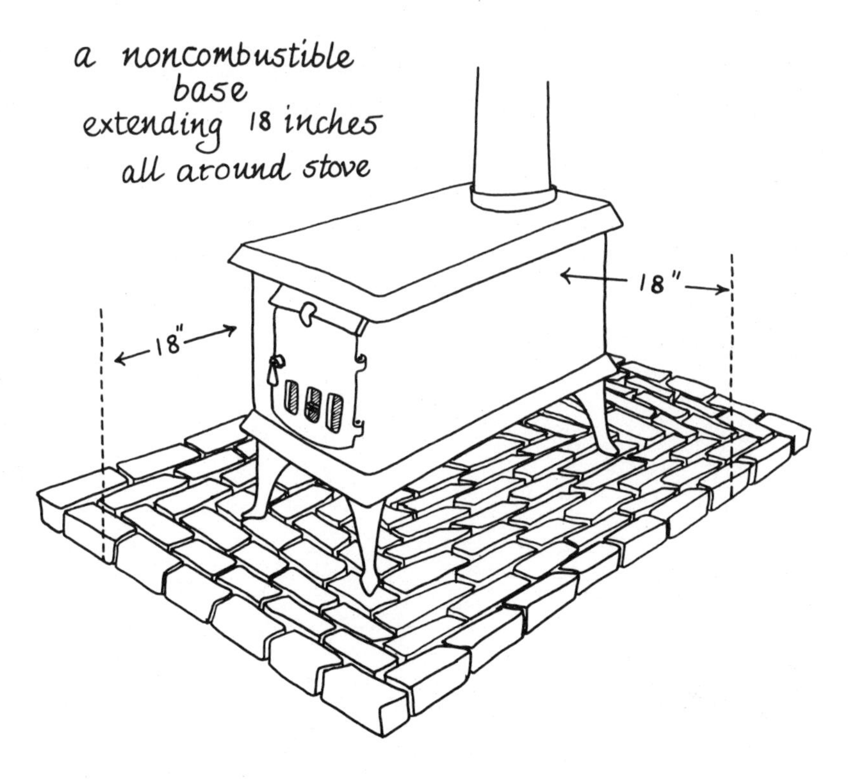

wood framing around the fireplace and any combustible wall is a yard from any radiating surface. The hearth must be of a noncombustible surface, but since heat rises, you can get by with just a layer or two of bricks over a wood floor. Just make sure that the combination of stove leg height plus height of the hearth material equals that 18-inch minimum above the wood! The hearth should extend at least 18 inches out around the stove on all sides—sparks do fly.

One respected fire safety organization says you can have your stove on little four-inch legs if there's a sheet of 24-gauge or thicker metal underneath for 18 inches all around the stove. I don't go along. Sheet metal can heat up and transfer that heat to a wooden floor beneath. Louise and I stick with the 18 inches in combination of leg length and cinder block, bricks, sand or whatever between the bottom of any stove and a wooden floor.

The same clearances apply for brick-over-frame construction used to make walls noncombustible; however, the stove is still 18 inches from the framing behind the brick and plaster. If

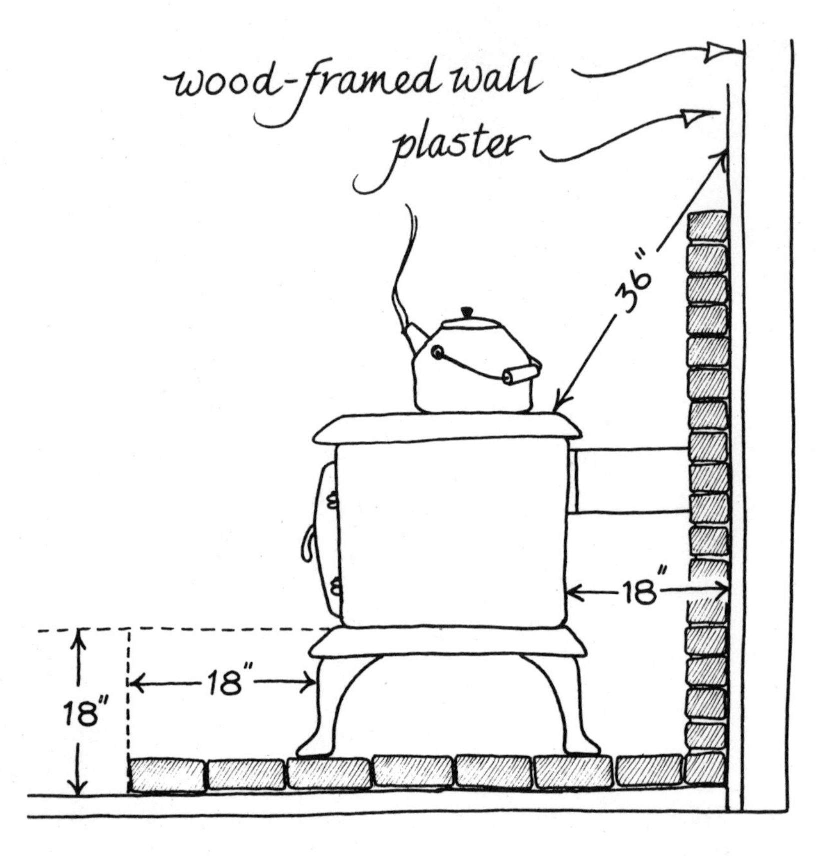

building a masonry backdrop is inconvenient, install a heat shield, a metal or asbestos/metal sheet six inches out from the combustible material in the wall.

With the barrier (and if a permanent installation, a layer of noncombustible insulation such as fiberglass between wall and board) the stove can be placed 18 inches from the wall. If you've any uncertainty about the safety of your own installation, and we haven't covered it here, get *Heat Producing Appliances Clearances* (NFPA no. 89M-1971) and *Using Coal and Wood Stoves Safely* (NFPA no. HS-8-1974) for a few dollars from the National Fire Protection Association, 470 Atlantic Avenue, Boston, MA 02210. And follow their advice. The publications are illustrated with photos of house fires caused by improperly placed wood heaters and captions give the numbers of people who died if you have any doubts.

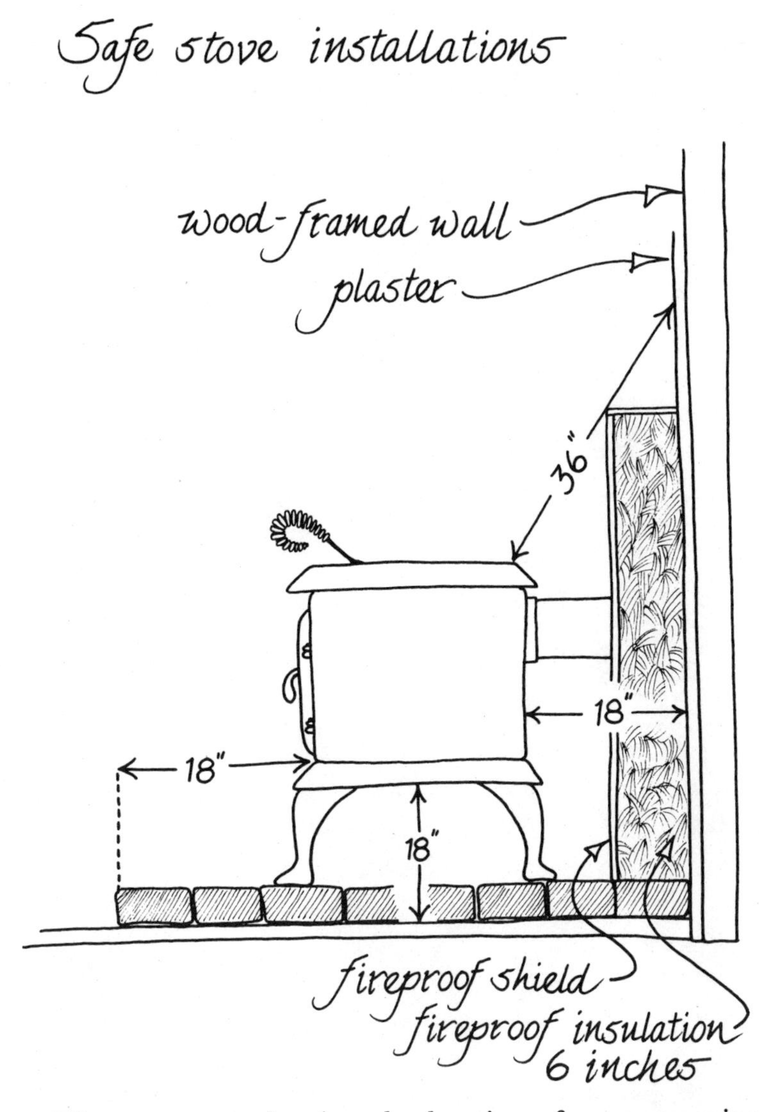

When you are planning the location of your stovepipe, remember, it heats up hottest at bends, both the elbow, if any, at the back of the stove where the pipe changes from horizontal to vertical, and again at any other directional changes along the pipe run to the flue. Hot gasses pour out of the stove and collide with the elbow, which forces them to change direction. Often the elbows will be the hottest part of your fire and may need special attention if they approach a combustible surface. Make the "never too hot to touch test" if in doubt.

When the main heating stove in the house was installed sometime back, we had an experienced mason erect a new flue for it. But the masonry hearth and backwall we built ourselves, as you can see. I bought a standard "cube" containing 544 bricks, selecting the kind with holes in them because they are lighter, easier to split, and cheaper. Splitting the bricks is done using a mason's chisel and hammer. I found that a single sharp crack with the hammer splits best and cleanest. We chose a corner location and a "sunburst" brick pattern for the hearth. After stapling a layer of asbestos-paper fireproofing to the floor, I used a length of string as a compass in

marking arcs for the courses and as a centerline in aligning the bricks. The bricks were positioned, and I dusted sand into the joints, filling out the final half-inch or more with mortar. A mason doubtless would not do it that way, but I'm no mason. Next, I outlined the fire-stop box and cut away plaster, lath and wall framing as necessary for it. The fire-stop box I constructed as explained in the chapter on flues. After lining the wall with the asbestos paper, I laid up the bricks, cheating just a little

by scribing lines on the wall so I'd know where each brick was to go (and that it was level and plumb without constantly checking a level). Metal wall ties were nailed to the studs and mortared between bricks. I finished up the fire-stop box assembly after the brickwork was finished, by loosely wrapping a six-inch stovepipe with asbestos paper, slipping it inside a seven-inch pipe, wrapping the seven-inch pipe with asbestos paper and slipping it into the eight-inch pipe in the fire-stop box.

The Damper

Many stoves come equipped with their own damper—that round plate a bit smaller than the stovepipe that revolves on a pin to regulate flow of air and smoke out of the stove and into the flue. With a stove lacking its own, you'll have to put one into the stovepipe. Recall what happened to us when I failed to put a damper on our first stove? Most commercial dampers come with a spring-loaded pin you can push in on the coiled-wire handle and the pin will come out of the guides in the plate. Then punch right-sized holes in opposite sides of the pipe (horizontally usually, though any way will work), put the pointed end of the pin in one hole, put the plate into the pipe, run the pin through guides in the plate, and out the other hole in the pipe. A bit of pressure and a twist on the handle to join the parts and you've a working damper.

Where you install the damper on the pipe is up to you—wherever it is handiest. You'll be using it a lot, opening it full to warm up the flue to get a good smoke-clearing draft when starting a fire and just before you add wood, and dozens of times a day as you use this rear air flow control in conjunction with the front draft controls to regulate the fire. So, place it in the pipe

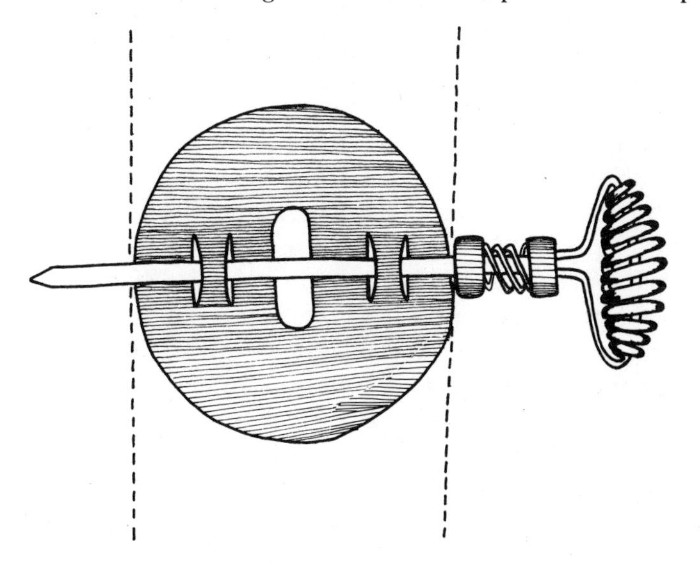

damper

where it will be handy when you are kneeling down in front of the stove with a load of wood in one arm.

Two Easy and Safe Installations

Probably the easiest and safest way to install a stove is to put it into or in front of an existing, safe and soundly mortared fireplace. Several traditional Franklin-style stoves are expressly designed with front-operated dampers for this application, and there are at least two modern designs on the market and more are assuredly coming. If the size of the hearth is big enough to provide the 18 inches floor space required all around, and wooden mantles, etc. are at least a yard above and to the sides of any part of the stove that could radiate heat, just put her in, connect the damper and run it into the flue. Stuff fiberglass insulation or stove putty around to seal it up and you're ready to fire up. If clearances are insufficient and/or if you want the stove well out into the room so it will radiate most efficiently, you can lay a fireproof foundation and run your pipe back and up into the flue or put on a fireplace cover. Many old homes have had the big fireplaces bricked up with only the stovepipe opening left. Check flue safety and clearances and go ahead and use it as is. You can make your own fireplace cover plate from steel, reinforced asbestos, brick or whatever. One firm, the Better 'n Ben's people listed at the end of the book, sells a ready-made cover plate and a square stove to go with it if you want. If you make your own, don't waste as much time as one fellow I read about who apparently never heard of pipe dampers. He put two holes in the cover, one for the pipe, the other with a door for his hand so he could operate the fireplace damper. (Though, this idea may not be as silly as it appeared to me at first. With this arrangement, you'd not have to be sure of a perfect seal around the borders of the cover and pipe hole to prevent warmth from escaping up the open flue.) If you have a flue in the wall, but no fireplace, the bricks will usually be covered by wallboard, lath and plaster or another decoration that could catch fire. You must remove the covering in an 18-inch diameter circle at minimum around where the stovepipe enters or construct a three-pipe firestop as explained above. If the heater is less than a yard away from the wall, you must also remove the combustible coverings to satisfy clearance or put up a brick or other shield.

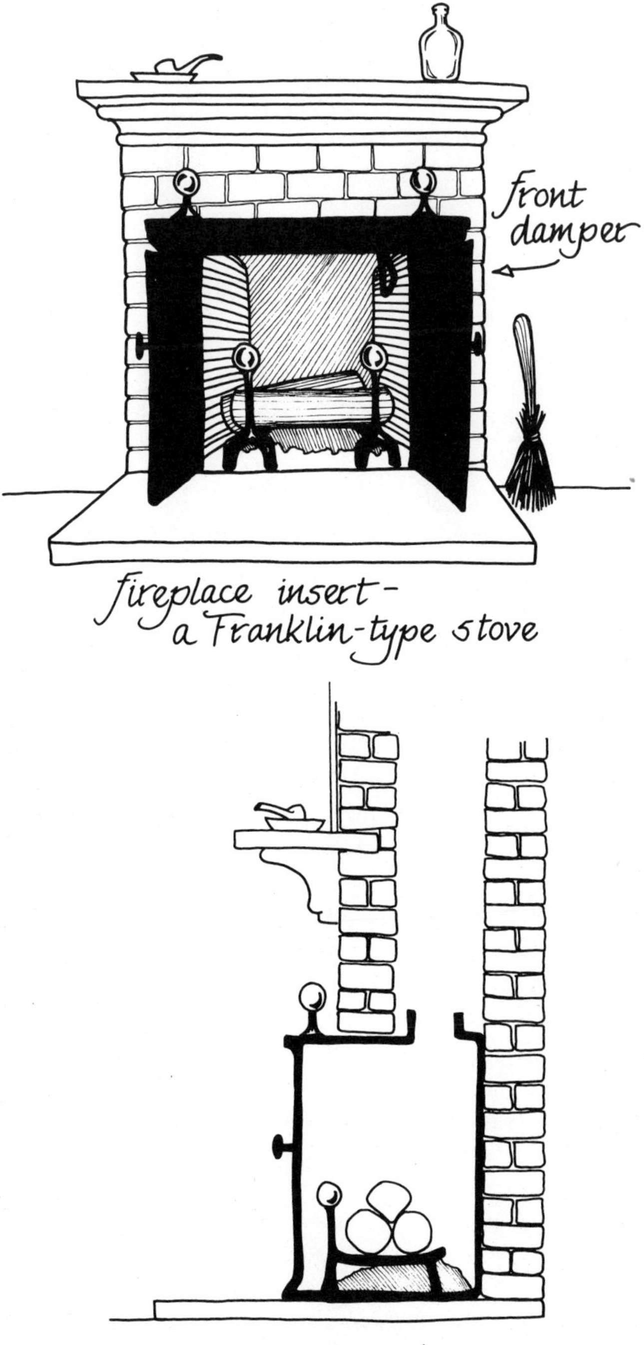
front damper
fireplace insert - a Franklin-type stove
side view

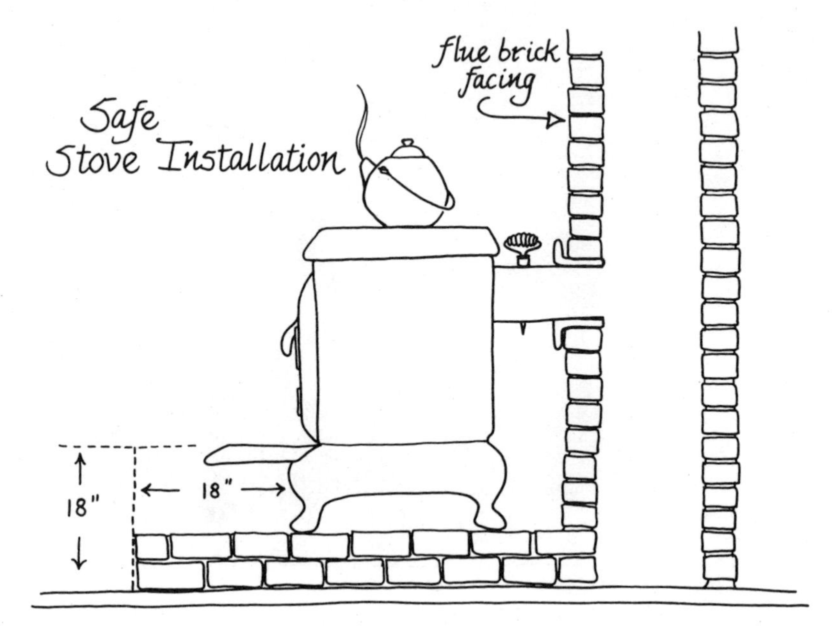

We explain how to cut a new opening in an existing flue and install a thimble in the account of our installation of the stove in the cellar. There the closest nonceramic material was in the floor beams above, and I just kept the stovepipe a good foot and a half below them.

By the way, let's touch again on the number of stoves you can vent into one flue. Basically, the combined area of all stovepipes exhausting into the flue should be a bit less than the area of the flue itself. Assuming that you have a standard foot-square flue interior, that means you could install seven six-inch pipes, six seven-inchers or five eight-inch pipes. That is, if all are operating all the time. Theoretically, you could pepper the flue with as many stove holes as you can get in without causing it to collapse, so long as only the above numbers were operating at one time. I'll have to repeat once again the admonition that your central heating should be on its own flue, and so should each fireplace. We live in a house where a central-heating system has peacefully shared a flue with a wood cookstove and two room heaters for the better part of a century, however.

Safe Operation

If you've made sure that the flues are safely constructed, then be just as sure the stoves, pipe connectors and your own operating techniques are, too. Wait till dark, then put a lit flashlight into your stove and close all doors, dampers and drafts. Check carefully to see if light is coming out anywhere it shouldn't. Caulk any cracks, say where castings fail to meet, with stove cement. You don't want an errant spark to come out where it isn't expected. It's a good idea to repeat this check periodically.

Be sure that all stovepipe connections are firmly mated. Loose connections can emit sparks, let pyroligneous acid out, and affect the draft. If your pipe runs any distance in the horizontal, wire it securely to the wall or ceiling. So long as the pipe is always running at a slightly upward slope as it moves away from the stove, you are burning open nonairtight fires, and you clean the pipe frequently, a well-secured pipe can be as long as your room is wide. I've seen pictures of old school rooms and churches with pipes 30 feet long and more. Again, be sure the pipe is secured well enough that a fire in it won't send the sections flying. I'd recommend securing each connection with three sheet

metal screws and wiring the center of each two-foot section of pipe to the ceiling to be on the safe side.

Controlling the Draft

Some stove/flue combinations are improperly matched—say a small, loosely constructed stove with a six-inch pipe venting into a warm flue that is 18 inches on a side. Theoretically, the flue could handle 18 stoves of that size and it can exert quite a strong pull on just one. This is usually the problem when stoves roar or overheat despite all you can do. Restricting the flue size at the chimney top might be feasible. Probably easier is to install

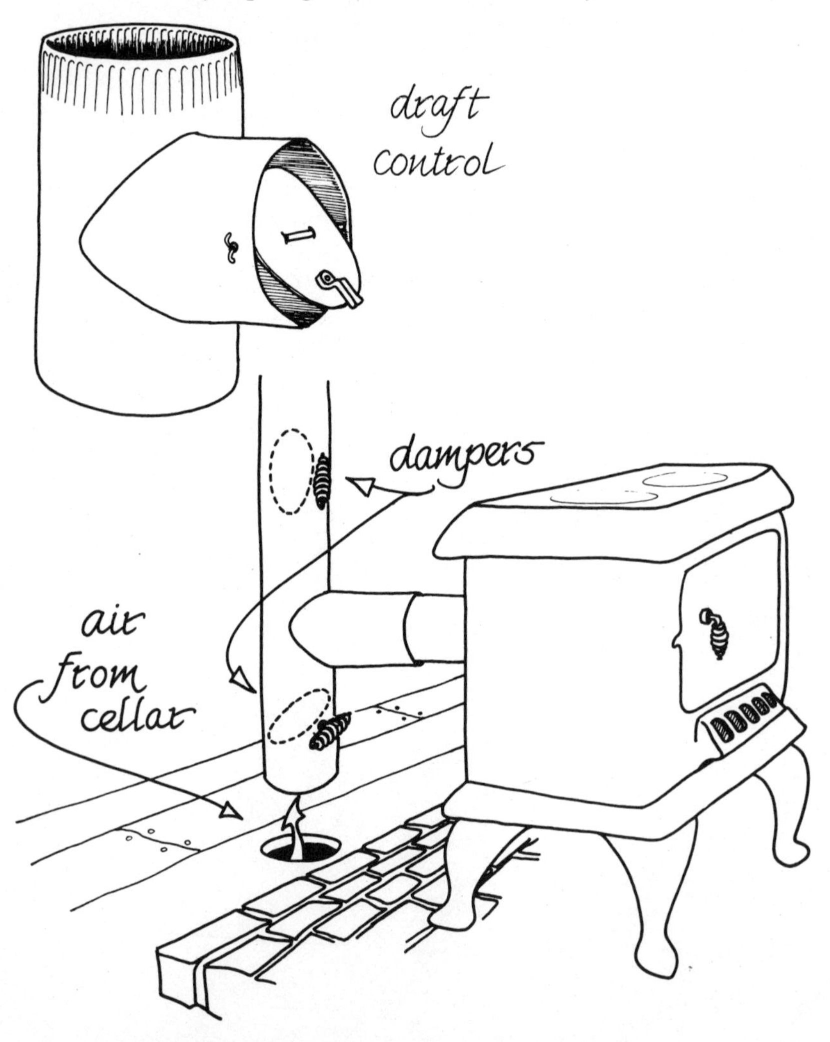

a draft control. These are devices that admit room air to the stovepipe to equalize draft. The automatic type that you find on most furnaces has a flap that is pulled open when the draft is more than the stove requires.

You can make your own draft control by installing a **T**-connector in the pipe and putting a damper in the open end. It can go on the pipe between stove and flue, above or below the stove damper wherever seems to work best. I've seen several ingenious uses of **T**'s to tame recalcitrant stoves. One, which will burn out the **T** in short order, placed it at the boot of the stove, the top of the **T** aiming vertically. A length of pipe with a damper was aimed down to a hole in the floor, the other, with a damper went on up to the flue. When the stove had more draft than it needed, the lower damper was opened and cellar air went up the chimney, curing the too hot fire but not causing drafts and loss of heat in the room. (I must admit, that same installation went into the drum stove we'll make up next.)

Make Your Own Stove

There is a lot of interest in homemade stoves these days and with the price the stores are getting for cast-iron models of any size, building your own makes sense. You surely aren't going to do your own iron casting, and unless you've a professional welding outfit and the skill to use it, boiler plate is pretty much out of the question. (Though plenty of small welding shops will cut and weld a stove of any design you come up with for a pretty reasonable fee; be sure they use nickel rods, though.) I suppose you could build yourself a brick stove on the Rumford design and install ready-made glass doors, though few will try.

The typical handy person (Louise being a steadier hand with a saw than I) is pretty much limited to sheet metal. Cutting metal of any thickness with common home power tools is tediously time-consuming, and you run through a lot of saw blades. Plus, bending and punching thick metal to make rivet joints takes heavy machine shop equipment. Making up a sheet metal stove from scratch is another job that a few may try, but only a few. Fortunately, industry provides the raw material for good stoves that you can get for just a few dollars—oil drums. There are three main sizes: the 55-gallon drum is the most common and easy to find and will heat a really big garage, workshop or cellar. A 30-gallon drum is harder to find, but is a size to fit in with the scale of a home. It also heats as much area as one of the small Scandinavian stoves, though it can't be made as airtight and fuel-efficient by any method I know. And the 15-gallon drum makes up into a nice one-room heater. Whatever the size, it's an "Alaskan" stove. The 49th state is still heated in large part by drum stoves.

Pick up your drum from almost any auto service station or factory, or get one from the drum cleaning service you'll find in any industrial area. Get a fixed-head drum with both ends crimped on. The kind with head held on by a thin strap could fall apart in time. The 30-gallon drum we have cost two dollars uncleaned and I probably could have gotten it for less if I'd

a museum dumb

haggled. You can make up your own design. Set the drum up on one end or have it horizontal. With some cutting and welding you can set one, two, or more drums in a stack, connected by heavy steel tubing. Put the stovepipe in at the back of the top drum and you've a modern version of the old sheet metal "dumb stoves" you see in museums. (Some of these had half-a-dozen or more tubes leading from one chamber to another.) The more metal you have heated by fire and smoke, the more efficient will be the stove.

Drum Stove Kits

More and more firms, large and small are offering drum stove kits consisting of a cast-iron or heavy metal door, legs and smoke boot or pipe attachment—all you need but the drum. Some include grates, which aren't really necessary in a drum stove. And, sad to say, some are poorly designed and made, probably by someone who jumped on the popularity of wood heat a bit faster than he should have. (This holds true for a lot of other items too—whole stoves included.) We ordered one kit that promised to let you see the fire. It consisted of a Pyrex pie plate and assorted flat metal stampings plus an indecipherable set of instructions. That was some time ago, and there are good kits available now, including ones with see-through doors. Most will cost under $50. Some of the stove foundries offer cast-iron ones, but we got a steel-fabricated model from Markade-Winnwood—address in the Suppliers list. It has a door that seals nearly airtight-shut, sturdy legs and a smoke boot with a built-in damper and is well worth the cost. In steady use, a drum will burn out in two years or so, then you just remove the kit parts and put them on a new stove. You may have to cut frozen-on stove bolts, but the rest of a good kit should last a good many years.

There are little sheet metal stoves appearing now in the hardware stores that sell for just a few dollars more than the materials for a drum stove. But in the ones I've seen, the metal is thinner than in oil drums, and the doors and smoke boots tinnier than in any of the better kits. I'd say, if you want a good stove, cheap, a drum and kit is your best bet. Let's build one.

Step-by-Step

The photo series shows the basic steps. Following kit directions, I cut door and smoke holes and drilled holes for various bolts. Then legs went on. Wanting a water-heating capability, I worked up a coil arrangement of copper tubing bent so the main run of tubing rests on the bottom of the drum, the ends exiting from holes in the back. Most drums suitable for stoves have a pair of holes, a bung and an air vent. If not, drill them where you wish. Our coil was designed to be pushed in through the narrow door opening. You could make up a coil that goes all around the inside of the drum and sort of screw it into the drum, I guess, but that's more work than I wanted to get into. As is, the coil is buried in the sand that covers the stove bottom up to the level of the door. A little pump, powered by a small electric drill and obtainable for under $10 in any hardware store, will keep water running from the storage tank when the stove is fired up and installed anyplace with 110-volt electricity. Next fall when it's time to cut the following year's wood supply, my woodchopping partners and I will take the stove out to the handmade shack that keeps us more or less dry at night. We'll try a gravity system, and see if heated water will naturally circulate as it does in the old, but better designed units used with coils in wood cookstoves. This may not work; it just might go to steam and blow us up without a mechanical circulator—we'll be careful at first. However, a small 12 volt DC pump from one of the mail-order surplus supply houses should work off the batteries of one of the trucks.

Into the stovepipe I put the drum oven, and installed the updraft system employing two dampers in a **T** located on the smoke boot that we illustrated earlier. This way, we can avoid the roaring blaze that a too-strong draft can cause in a nonairtight stove, plus by using the damper in the boot together with the two dampers in the **T**, we can regulate the amount of heat getting to the stove; close the bottom damper in the pipe and open the other two and you've got biscuit-cooking heat in a minute. To cool the oven, put in a cold brick or open the bottom damper and regulate the other two depending on the fire's heat.

Total cost was well under $100, $35 for the kit, $2 for the barrel, $3 for the little pot rest that goes on top of the barrel, making a flat cooking surface, around $25 for the drum oven, another $2 or $3 for the water storage tank, about $10 for the pump which has many other uses, a few more for the tubing and fittings and we already had the electric drill and soldering equipment. Hard to beat for a complete space heating, water heating, top and oven cooking unit, wouldn't you say?

One caution, though. Industrial steel drums are used to transport some pretty noxious substances. Ours held one of those herbicides they use to keep brush down along roadsides, to defoliate Southeast Asia and like acts of environmental idiocy. Not wanting any living thing to be exposed to the possibility of breathing the stuff, I put the stove in the pickup, drove to the woodlot, fired it up and stayed downwind till all the paint had burned off the outside and the smoke contained not a whiff of chemical odor. Hopefully, I didn't gas any owls. And that's where the stove is now, in the sugar shack out at the woodlot, with a rust-preventing coat of stove black over all. When we go woodcutting come fall and during the spring maple sugaring weeks, it'll be good to have a warm place to sleep, hot water to wash up in and oven-baked biscuits to go along with the ham and

Assembling a stove from a kit and an old drum is not difficult. First, mark the outlines of the kit parts for cutting and the locations of holes for drilling. When cutting the smoke hole and loading door, use a saber saw (with a blade designed for cutting metal) for the straight runs and a cold chisel for the corners and curves. The final

eggs in the morning. The only disadvantage of this design compared to the little potbelly it replaced is the small flat-cooking surface, capable of holding just one pot or skillet at a time. (Without the pot rest on this design, there'd be no skillet cooking at all.) The little drum oven more than makes up for it, though. But if you don't want to bother with the accessories and will be cooking a lot, build your stove up on end and exit the pipe out the back, not the top. The top will have a slight bump to it, but a good cast-iron skillet and your Dutch ovens won't mind.

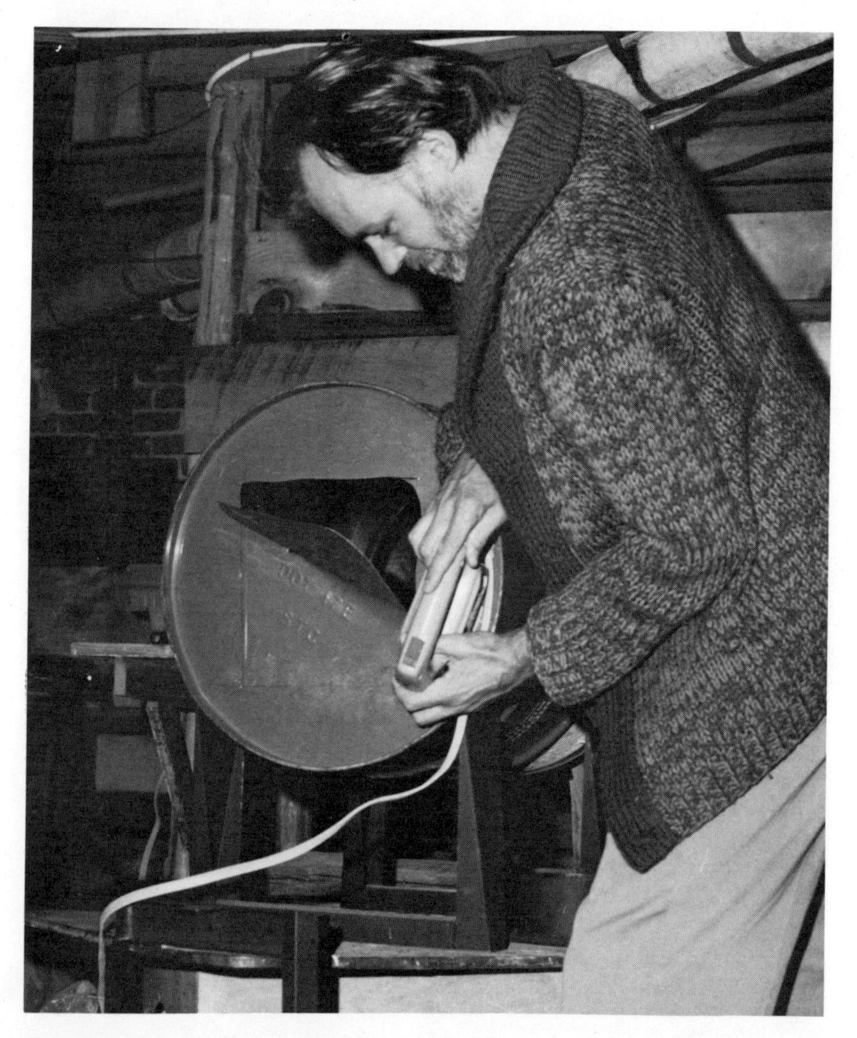

assembly step is to bolt the kit parts to the stove. Haul the stove outdoors–to our woodlot, in this case–and burn out the drum to remove paint and residues of whatever the drum originally contained. In less than an hour of burning "squaw

wood," dry twigs and fallen limbs that burn hot and fast, I had the noxious smell fired away. Back at the house, I used a wire brush and electric drill to scour off the remains of the drum's paint, then coated my new stove with stove black. Though a

messy job, a good application annually will protect the stove from rust. I added a couple of special accessories to my drum stove, a stovepipe oven that is becoming a commonly available item, and a water-heating system that is of my own construction.

Firing Up the Stove

Before going any further, let's fire up your new stove. You'll want a handful of small splinters, as dry as you can make them, a half-dozen kindling sticks about an inch thick and three or four medium-sized pieces three to four inches through—"quarter-splits," or your typical firewood-size log split into four parts. Then, you will need a supply of big logs, whole limb pieces and "half-splits" of larger logs, eight inches through and larger. Be sure the wood is warm. Bring it in the day before to let it lose its chill and have time for surface moisture to evaporate. Then, open the stove door and damper all the way. This is to let the flue warm. Remember, cold flues will chill the first smoke, sending it down and back into the room; oxygen won't be pulled through the wood, and instead of a well-fired stove, you'll have a smudge pot. Cold flues don't draw, as they say, and they attract creosote, so let yours warm up.

The firing process is similar for any stove with a grate or wood basket, andirons or whatever to hold wood up off the stove floor. This includes the circulators, Franklins and other open stoves, potbellies and their kin. Loosely crumple several sheets of paper between the andirons or into the firebox and lay on a couple of handfuls of splinters. Any which way is the best arrangement so long as there is plenty of space between each. Now put on the kindling, in a crisscross pattern in square stoves, a tipi shape in the high railroad types and potbellies, longwise in a series of **X**'s in the long logburners. Some people lay in the larger sticks now; I like to light up, let the paper collapse and be sure the kindling has caught well first. With cold or wet wood and a really cold flue on a really cold day it sometimes takes several starts to get up a good blaze. There's no need to keep pulling out partly scorched wood to re-lay the paper and kindling.

kindling arrangement in:

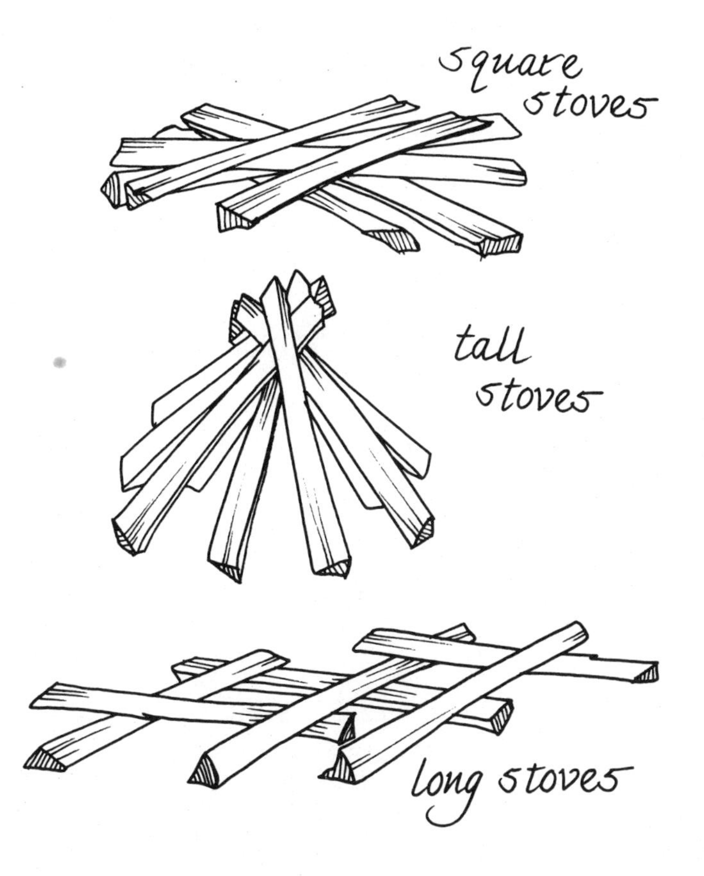

Once kindling is caught, I put in as many quarter-splits as I can get on in one layer, leaving just a crack between each piece. This is Louise's own magic fire-starting principle; before she figured it out I was taking ages to get a fire going. She's still better at it than I. The narrow opening between logs increases the velocity of hot air flowing up from underneath, in effect "blowing" on the fire to increase oxygen supply at the kindling point of the wood. With split wood there are little splinters sticking out, and splits catch faster than whole logs.

When the first layer of quarter-splits is going well, I put in some larger pieces of split wood, another single layer to fill the firebox from side to side. In a half hour or so a good bed of coals will be established and you can fill the stove and close the damper halfway. In perhaps an hour more the wood will be hot and you have yourself a stove fire. Here on in it's between you and your stove to figure out what combination of damper setting and air control will give you the heat you need.

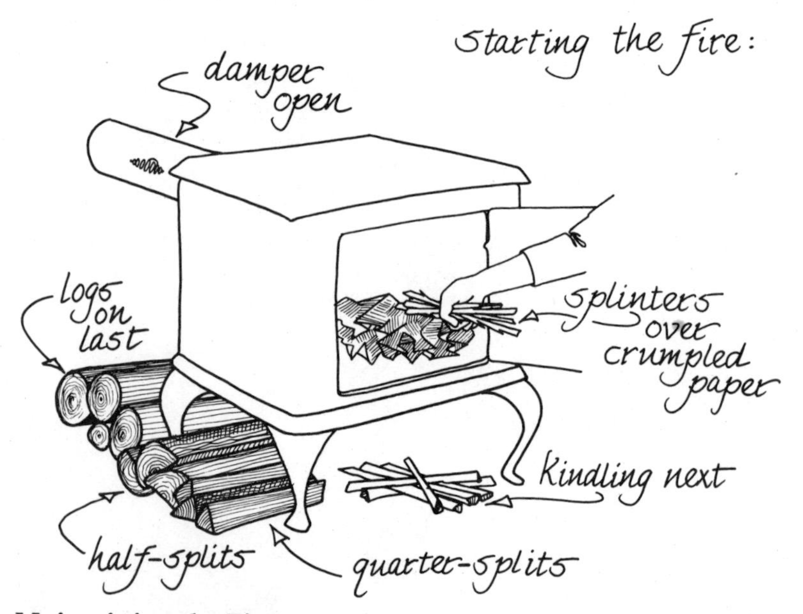

Maintaining the Fire

With any open stove such as a Franklin, the best fire is the fireplace-type ash bank that we'll cover in detail in the following chapter. But most closed stoves have two air entries or draft controls. Some have three, the third a skirt at the very bottom of the

firebox tha
the fire. Bu
with grates)
one is to adm
bon in the co
combustion of
every stove/flue
wood and even v
need more warm
relatively warm d
wood.

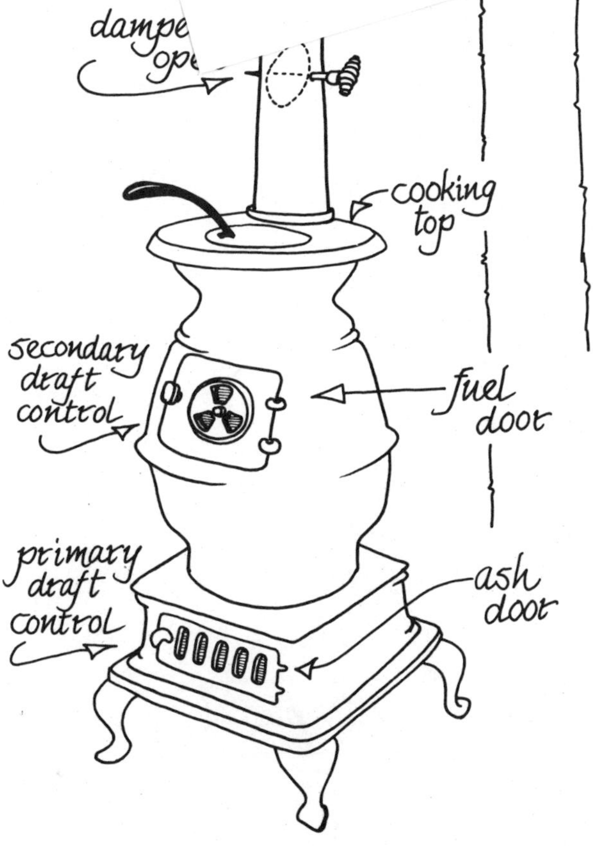

p a stove and opening the
ake for the quickest starts. It's
Then when a good coal bed is
n often be closed completely so coals
top opened to one or another degree to
mbustion of the gasses. But the ideal heating
per completely, or almost completely closed, the
up with just enough air entering at each level to
the degree of heat you want. The more heat, the more
u let in, the wider the damper opening and the more often
u reload with wood.

In almost any stove, reloading a well-started fire is simply a matter of piling in more logs. Unless the wood is wet, in which case you'll often need Louise's magic spacing to get it going, a fire will take wood any way it gets it.

Firing the Grateless Stoves

Many of the imports lack grates or andirons, and firing them the first time is a different proposition. Open damper and door to let the flue warm as with any new fire. Lacking grates to keep the wood up (in one of the long baffled models) I like to put a short but fat log on crosswise about two inches back into the stove. Crumpled paper goes in back, splinters are scattered on top, and the kindling laid in so the front of the sticks rest on the crosswise log. That way, plenty of air gets to the wood. When kindling is going, in go the quarter-splits, and then the big logs. After a day or so of burning you'll have enough ash that you can rake an ash mound replacement for the starter log up to the front of the stove. Then just keep putting in the wood.

With the upright imports, the Scandinavian combis and a few of the Central European iron and ceramic stoves with or without grates you can use the tipi arrangement of paper, kindling and logs to get the fire started. In time the tipi will collapse into coals and you can continue adding wood in tipi fashion or as it happens to fall. Many of these stoves were designed for coal or

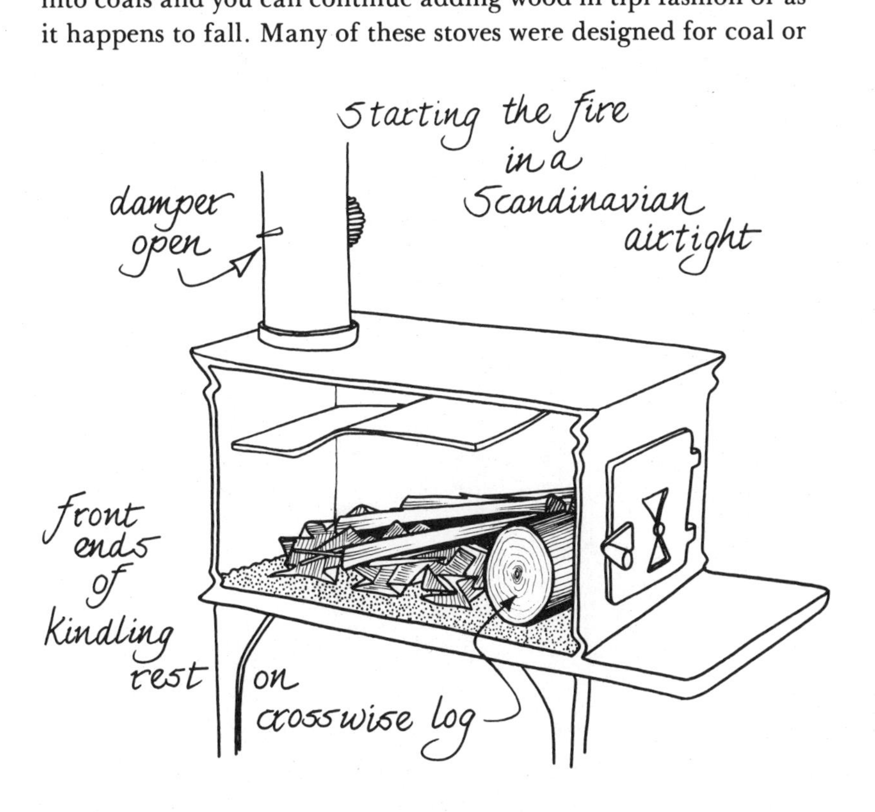

coke burning; in fact most stoves that are considerably taller than they are long are coal or coke burners.

The cylindrical or high square-topped coal stove has a firebox no more than a foot square and if you try to set big logs in on their ends they will often snuff out the fire. You'll have to split virtually all logs more than six inches through to keep a fire—something to think about before taking one on. Folks I know who do use this type of stove—the tall, slim design goes beautifully in many rooms—find that the best fuel is slabwood, the thin pieces cut off to square logs up for cutting into lumber at the sawmill. More on slabs when we get to "Getting in the Wood."

With stoves such as our Combi which offers the open fire of a Franklin with a well-sealed door of the other super-efficient air-tights, you have a choice of fire design. A Norwegian friend starts her fires in the tipi form and keeps them that way. The coals and ashes are maintained in a cone shape in the center of the firebox and logs are laid up against the pile at the sides and pushed down

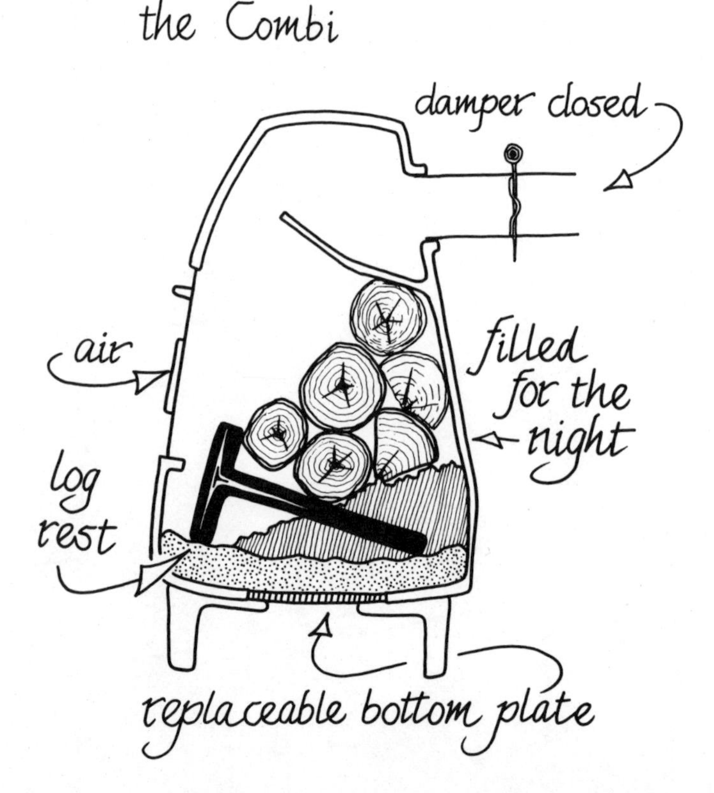

into the ash at back. Dead ashes gradually work their way out to the front where they are removed. Louise and I hark back to ancestors from a bit south of Norway, and like our cheery open British fire on its andirons or log rests. We couldn't find a set small enough for the Combi, so I got the smallest set of log rests I could find and had a machine shop saw one in half and cut down the legs. They rest on the bottom of the stove, angling down toward the back. During the day we keep the banked fire described in the fireplace section. At night, a half-a-dozen or so logs are stacked in against the bed of ash and coals and the door is closed, damper shut down and draft opened depending on the temperature outside and degree of warmth needed. Settings will differ with each installation and you can learn only by trial and error.

Just don't shut any airtight up completely (or almost completely—all dampers are designed to let some air through). One neighbor closed up his American-made airtight stove the first night he got it, came down the next morning and found a cold room and the stove full of perfectly distilled charcoal! Plus, I'll bet, a brand-new layer of creosote on the flue liner.

Pajama time for the kids means we can bank the fire and close the big stove for the night. With a full load of logs and the damper almost fully closed, the stove will keep the house comfortably warm for sleeping all night long.

Secondary Combustion Chambers

Now, we obviously advocate the low flame, good-heating fire. Indeed, when we get to fireplaces, where you can really inspect a fire as it burns, we'll see that the best heating fire is one with barely any visible flame. But some folks will want to install the bigger potbelly types and homemade barrel stoves in garages or drafty workshops where the gentle glow of an efficient airtight-style stove fire just won't keep you warm. What you want is a real barn-burner, a fire with a draft that will suck a sheet of paper in and have the front half burnt before the tail end gets through the stove door. Here, you open everything up wide and keep pitching in logs to get the maximum safe amount of heat radiating out.

This will be a fire with an unavoidably high flame, the sort that potbelly stoves were built for. The high top pretty well contains the flame from the fire low down in the "belly." (For high-firing, build your homemade drum stove standing upright too.) And with any stove used for high-flame-type fires, you may want to add on a secondary combustion chamber—an extra chamber where the flame can be contained and a bit of extra oxygen introduced so the gasses expelled from the stove itself by the high flame can burn more completely. Often the stovepipe or flue acts as an inadvertent secondary combustion chamber, as too-high flames lick out of the stove. This is dangerous. Not even heavy-duty stovepipe is built to hold a fire and live flame in the flue can ignite built-up creosote. So add on the secondary chamber. A steel drum with stovepipe-sized tubes of the thickest gauge steel your machine shop can bend, nickel-welded onto openings front and back makes a good secondary chamber. In the end nearest the stove, drill a one-inch-diameter hole. Take a three-inch-diameter circle of metal, cut a notch an inch deep and wide in one side and drill a stove-bolt-sized hole in the center of the disc and another just to one side of the hole in the drum. Turn one of the disc's end points up for a handle, attach it, and you have a secondary draft control. Move the disc so the notch uncovers as much or little of the hole as you want. Depending on your fire, you may want a bigger opening, or two, one on each side. Attach the drum to the stove and you're in business. For stoves venting from the back, the secondary chamber will probably have the attachment tubing put on the ends. For a potbelly or other stove that vents from the top, you may want to put tubes on the barrel's sides, the draft control on the end facing the front of the stove.

For horizontally designed—or any—home-built drum stoves, you can attach a second drum over the first as in making a modern "dumb stove." I'd support the top drum with an angle-iron bracket at the rear, have the connecting tube or tubes toward the front, and the secondary draft in the front end of the top drum. Then keep the hot fire well back in the bottom drum, vent the top one from the back side, and you will be forcing the flame into a C-shape as it travels through both barrels. This arrangement and a high fire can singe your whiskers from six feet away.

variations

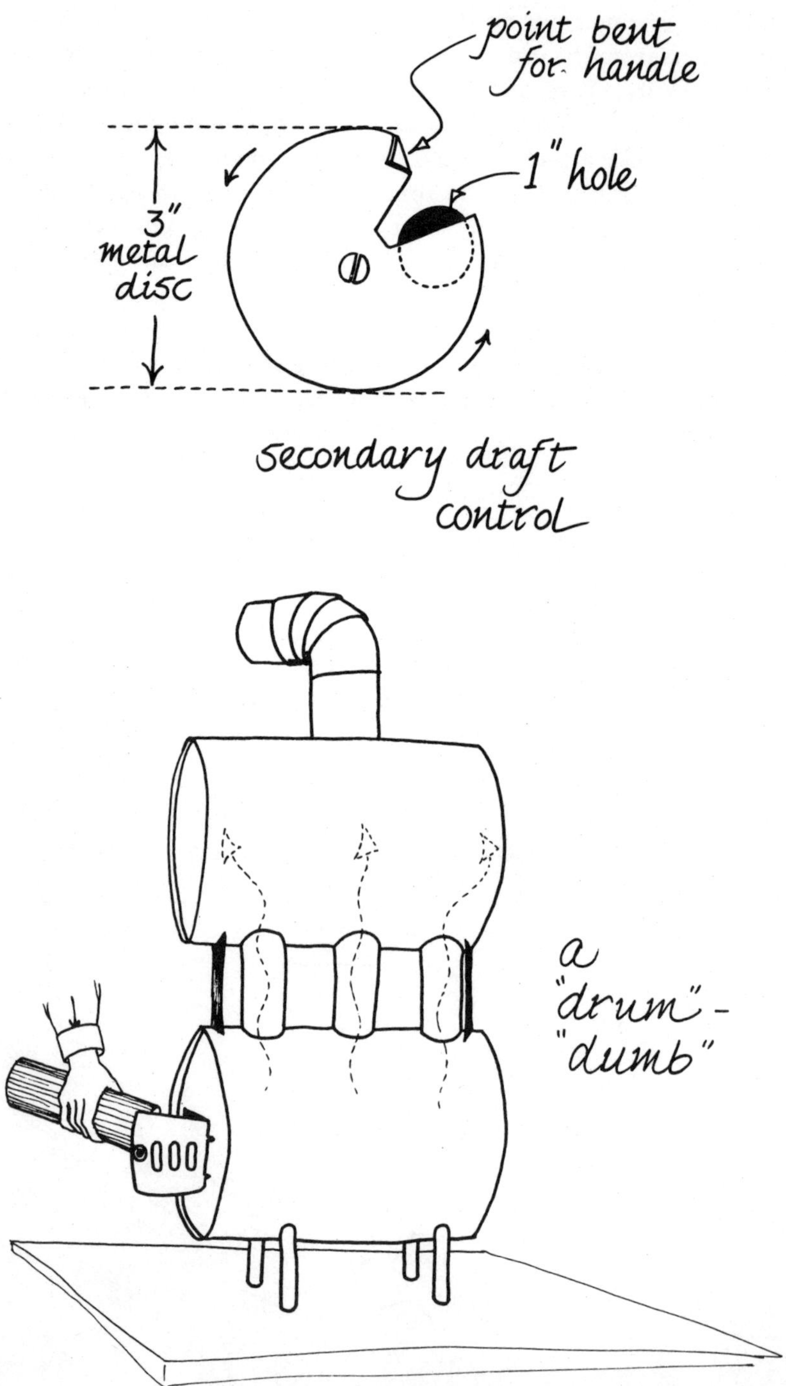
point bent for handle
1" hole
3" metal disc
secondary draft control
a "drum"-"dumb"

Fire and Safety

A hot fire is fine so long as the live coal bed doesn't get thick enough to generate enough heat that it will burn through the stove, warp plates, ignite creosote in the pipe or kindle any near-by flammable material. If you need a hot fire, be sure to set the stove up on several layers of brick or other noncombustible bases and keep flammable materials of any sort a yard or more away from it in all dimensions. Also, if you are running any length of stovepipe inside, secure it extra well and clean it frequently. You can smell an overheated pipe or overfired stove; it has that hot iron sort of industrial smell that Louise and I first noticed at our initial near-disaster back on the farm. To keep the stove safe, keep the fire within bounds by closing damper and door to restrict air flow and don't let the coal bed build up by continually adding new wood. Let coals die partway down once in a while.

When maple sugaring in late winter, cutting wood in fall and during other cold-season activities out on the woodlot, we've found that each time someone comes into the shack to warm up, he fills up the stove. With a half-a-dozen people tossing in wood, that coal bed can get up—perhaps when no one happens to be around to watch. It's best to let a single individual run the fire if possible, at least until everyone is accustomed to the stove. And, as we said in the beginning, never use cold water to calm an overheated stove. Hot water (or the boiling cider I used those many years ago) isn't the best answer either. Close off the air supply and flue and let the fire burn itself out. Save your water to cool down any walls or other flammables that may have been brought close to tinder point by a too-hot stove.

While we're ending a fire safely, I should inject a safety rule about *starting* one. *Never* use gasoline, charcoal starter or whatever to get an inside fire going. They are OK for the outdoor barbecue only. *Never* use them inside. *And particularly never* add, say a squirt of lighter fluid, to an ornery, inside smouldering fire that won't seem to get going. It can flash back on you or hit the fire, vaporize instantly in the heat and ignite with a really big bang. It's dangerous to have any flammable liquid in the same room with a wood fire, in my opinion. I don't even like those Cape Cod fire lighters that are becoming popular again. They have a chunk of porous rock on a handle that sat in a pot of whale oil in the old days. Now it's kerosene or charcoal lighter. The stone soaks up the stuff, you build the fire over it and light the stone, which burns off the liquid slowly, giving the fire a slow, steady start. You take the lighter out when the fire is cold. Well, it does make things easy and the little brass gadgets are an attractive ornament alongside stove or fireplace. But if an errant spark happened to pop out as you were adding wood and arch over into the pot or if a child, temporarily transmogrified into a P-51 Mustang on a combat mission happened to zoom by and kick the pot into your Franklin. . . . Well, I don't like them! Besides, a proper, "for-heat" wood fire only gets lit once a season. You just keep adding logs. Who wants a lighter buried in the ash bank all winter? It would burn up, most likely. Or who wants to pick it out, sizzling hot when you are removing ashes?

Ashes, Soot and Accessories

As for the ash, you will have some 50 to 60 pounds of it to deal with per cord of wood you burn. You'll want a coal scuttle and a long-handled flat shovel like you find in fireplace tool sets. With stoves having grates, you just shovel it out from under anytime it builds up to a level near the bottom of the grate. Most grated stoves are designed to have air coming up from the bottom, and if you clog them, the fire won't burn properly. Some models, airtights and the circulators mainly, have ash drawers you just pull out. With a going fire, the ash will be hot. It is also light enough that it flies easily, so shovel and dump gingerly. With any stove in constant use, the dustpan and whisk broom are as important accessories as the tongs and poker—more important really.

In grateless stoves, you really need an ash rake, a piece of sheet metal maybe four inches wide and two inches high attached in a **T** to a rod that's a handhold longer than your stove is deep. You may have to make your own the right size, especially for a big homemade drum stove. The foundries making cast-iron kitchen ranges supply rakes that will suffice for all but the largest stoves. With a really big stove, you may be able to use a rake from an old coal furnace.

With tall fires in upright stoves, the pressure from constantly

added wood and a little rake work do the moving. In stoves that burn wood lengthwise, "like a cigarette," you'll be pulling coals from the last-burned back end of the logs up to the front each morning and when you stoke up for the night. Just shovel the excess ash out into the scuttle.

If you use old-fashioned wooden matches, get an old-fashioned matchbox holder and nail it high on the wall out of reach of children, mice, and sparks, and heat from the fire. And while we're on fire-tending accessories, one item you don't need is one of those gadgets advertised to "turn newspapers into fireplace logs." For one thing, most folks come equipped with a dandy set of paper rollers, one at the end of each arm. The crank or roller devices are not only unnecessary, they take longer and don't work as well as your own two hands. The rollers come with several dozen lengths of twist-ties for holding rolled paper logs together. They sell extras for something approaching a dollar a dozen. The ties are nothing but cut-up lengths of the same wire and paper "twistems" we use to hold tomato plants to stakes in the garden. If you must burn newspaper (rather than recycling it into good things such as this book) you can buy a quarter-mile of uncut

twistem for the same price as you'd pay for a set or two of "official" paper-log ties.

And then, newspapers don't really burn very well unless they are loosely rolled so they go up in seconds. You must have a good bed of wood coals to burn well-rolled paper logs, and once they are burned to charcoal, you must crush the blackened, half-burned logs down into the coals to force them to finish combustion. I'm told that if you roll logs good and tight, but with a tubular opening through the center, then soak and dry each one, the paper will compress into a fair approximation of wood. We've never tried it and probably never will so long as our town has a paper recycling center.

If you do burn paper, stick to newsprint or plain kraft paper—the brown stuff shopping bags are made of. Most magazines are printed on coated stock which is covered with clay and other finishing agents that won't burn, and even if you poke them enough to get partial combustion, they'll fill your ash pit with chips and flakes of unburnt paper.

Reclaiming Heat From the Stovepipe

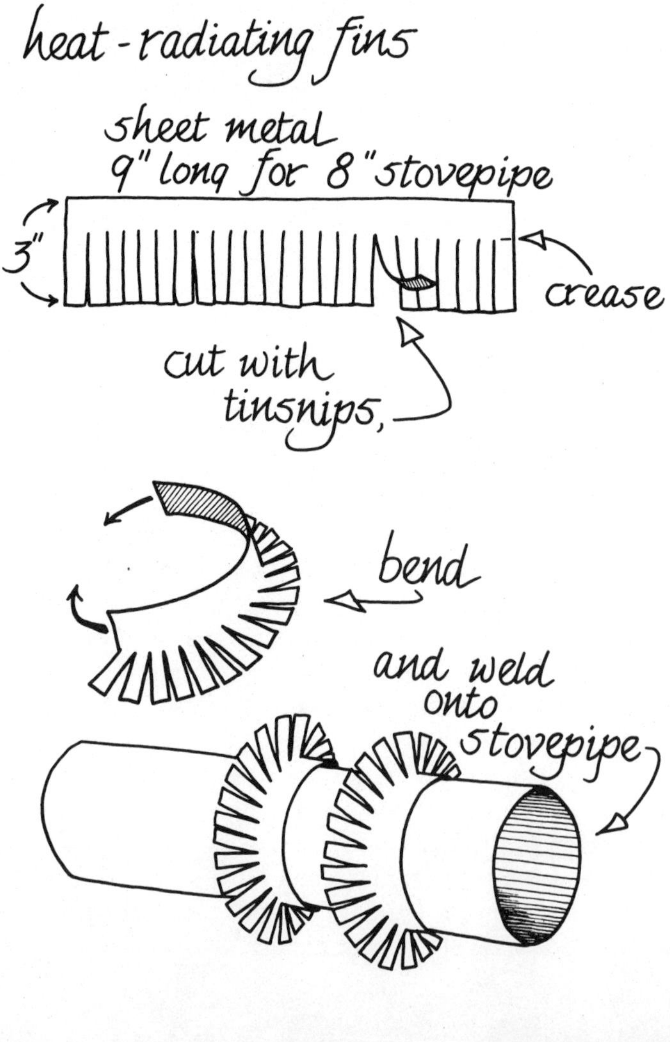

The smoke and gasses leaving a stove can reach 1,000 plus degrees F. and that heat should be salvaged if possible. The extra-long stovepipe just radiating heat into the room is a good measure, as mentioned earlier, though it must be kept clean and well guyed. Several firms are making safer kinds of pipe heat reclaimers now, and more are sure to come on the market. The simplest does nothing but increase the heat-radiating area of exposed stovepipe by wrapping it in strips of aluminum crinkled into a continuous **M**-shape. You can buy a set for a ridiculously high price, or you can make your own from strips of aluminum flashing obtainable at any hardware store. More effective but none too attractive would be to weld on fins per the illustration.

There are two basic designs of mechanical heat reclaimers on the market now (alluded to earlier). The most sophisticated uses a space-age heat pipe that employs a fluid circulating inside a closed tube to pull heat out of the flue and into the room. One unit I know, combines these pipes with a fan in a housing that fits right into the stovepipe. The other design mentioned earlier runs tubes through the flue gasses and blows air through them. Electricity used by the fans is negligible and all of them increase the heat output of your wood pile.

As of this writing, these gadgets all cost over $100, and in my opinion are greatly overpriced. Perhaps by the time you read this competition has developed to the point that prices are more reasonable. I hope so. If not, you might consider my favorite stovepipe heat exchanger, the drum oven we put on the oil drum stove. It even has a heat indicator in the door. Stick it in the stovepipe, put a damper above and below. Leave the door open

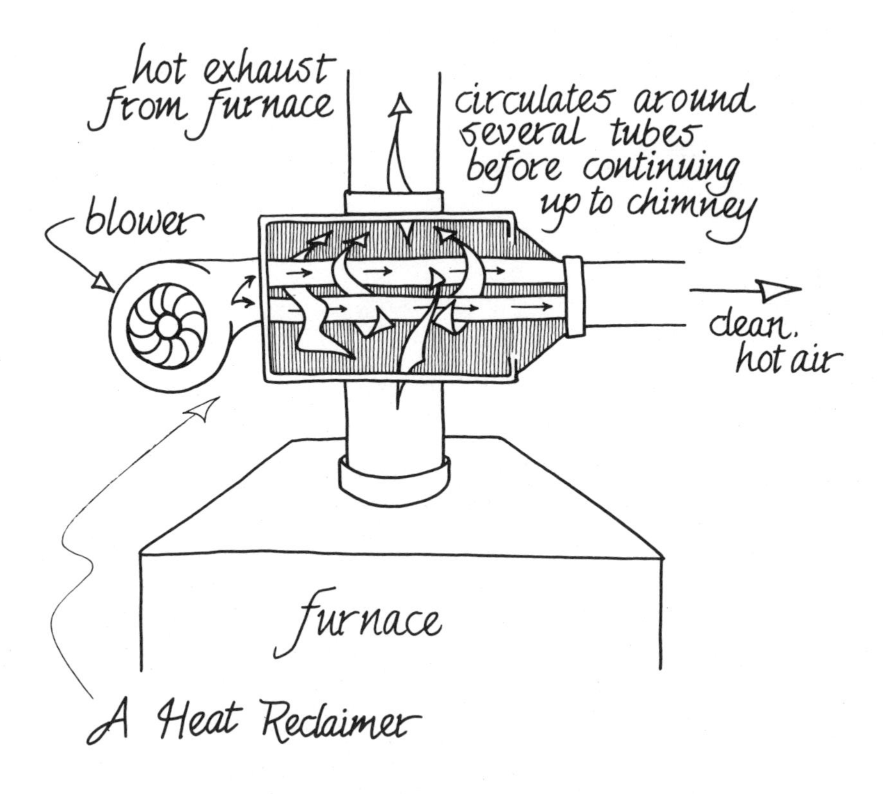

and you've an excellent heat reclaimer. There is even a scraper inside the hollow donut of the outer and inner liners. (Regular reclaimers have one too, I must admit.) Before firing up each morning, you just give the oven scraper a couple of turns to dislodge any soot that gathered overnight. And with a bit of experimenting, my version turns out a great stew, pie, or loaf of bread. Less than half the cost of the fancier heat reclaimers that can't so much as boil water.

I've never seen it done, but someone surely will someday wrap copper tubing around a stovepipe and run water through to heat it. I'd put a length of larger diameter pipe around the device with caps at each end to keep all the heat in. I don't know how much tubing would be needed or how fast or slow the water would be pumped to give it enough time to heat, but that's a question easily solved by trial and error. If you get it figured out first, let us know, won't you?

Central Wood Heat

I guess the ultimate form of wood heat is wood-fired central heating. That is, forced or gravity hot air, circulating hot water, or steam systems. Till the oil burner was put in, the house

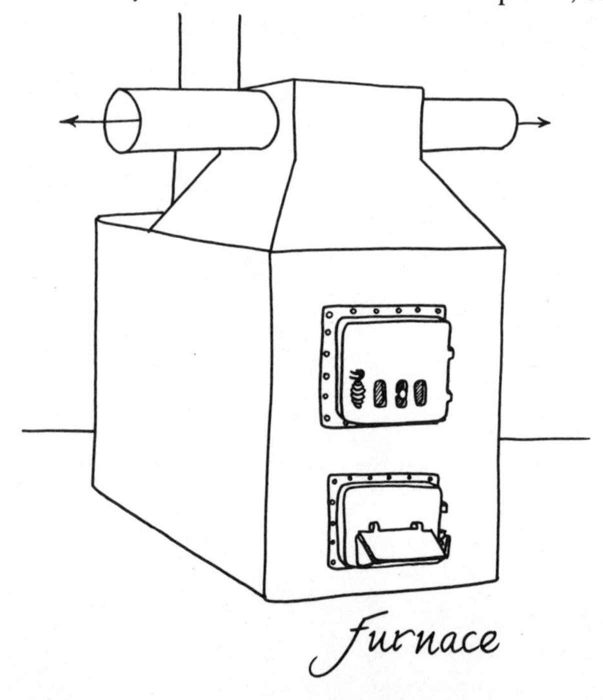

furnace

back on the farm was "centrally" heated with wood. In the cellar was a huge old sheet iron and firebrick wood-burning furnace that exhausted through stovepipe running half the length of the cellar and into a flue built a yard out from the house (by folks who'd lost one house to a flue fire and weren't about to lose another). The whole furnace was surrounded by a galvanized-tin skin forming an airspace a few inches deep all around. The top of the skin opened into a big wrought-iron grate that was still in the floor last time I looked. A fan rigged in back pushed air through a baffle system so it circulated all around the hot iron before pouring up into the house. Gratings in the first floor ceiling let warm air into the upstairs, and most of the house stayed warm enough, though the folks who last used the furnace claim you could toast marshmallows over the iron grating when she was really fired up.

Plenty of older homes have old coal burners that have been converted to oil, simply by putting an oil burner in and closing down the draft openings. Few of these have a large enough fuel box or the properly sized venting system to use wood and I'd not try reconverting a coal burner to something it never was. Such a

"Central" heating

furnace might adapt to burning charcoal, though, particularly if it is an old model designed to use bulky, clinker-producing bituminous coal. Few folks will have the necessary amount of charcoal available, cheap, but if you do, get ahold of the oldest employee at your local coal dealer (if he's still in business) and ask him to come out and see what can be done with the furnace.

There are at least two Canadian companies and one U.S. firm making wood or wood/oil furnaces incorporating modern furnace design improvements. You'll find them in the list of sources at the end of the book. And by the time you read this there will be more available, including some from Europe. There are also a number of small machine shop operators in our neighborhood who are using good old Yankee ingenuity to come up with designs of their own. You may find such an entrepreneur in your area. Largely because demand is as yet fairly low, you'll find that these units come for quite a lot of money. But they can be easily worth it especially if you have your own free and essentially unlimited wood supply.

Most modern wood furnaces are two-stage burners. That is, they have a lower combustion chamber that burns the charcoal and another where most flue gasses are burned. Fuel efficiency is

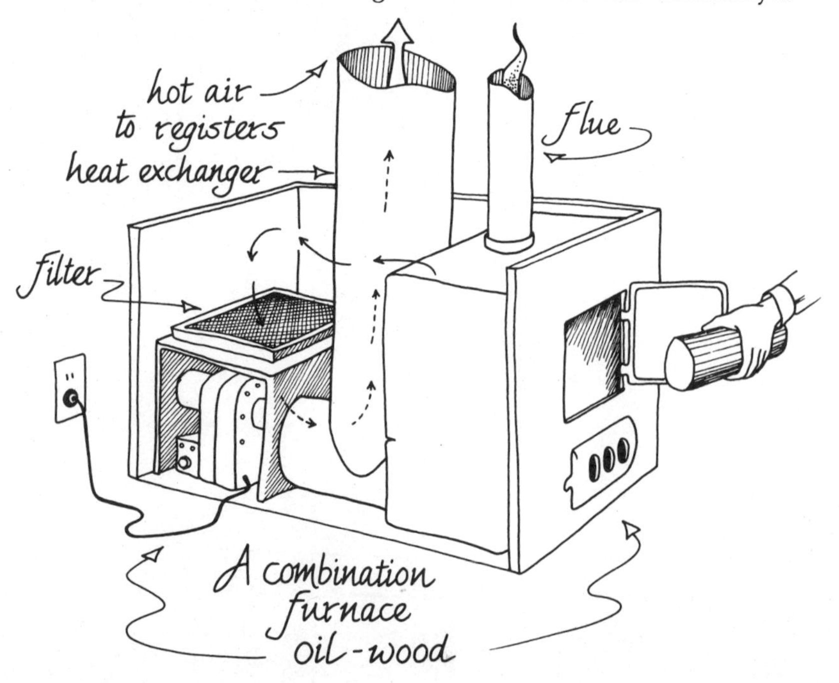

greater than with most space-heating stoves or any fireplace, and creosote buildup is lessened. Heat from both chambers is passed through a heat exchanger to a hot air plenum or a water or steam boiler. They all have automatic draft controls, some electric, others (simpler but more foolproof) operate on the bimetal strip principle used in some airtight stoves.

The most flexible (and the most expensive) models have a second heating chamber with an oil burner. If the wood fire gets too low, the oil kicks on. Dandy convenience, particularly if you plan to be away from home at all during freezing weather; with only the wood heat, you'd have to drain water pipes and remove anything that might break or be harmed by freezing such as all the food put up in glass jars. (Though some wood-burning folks arrange little electric heaters strategically in the house and set them to keep things just tepid while they are away.)

The big central furnace attaches to any standard air, water or steam system, and unless you know more about modern heating plants than I do, you'd best have the installation done by a heating contractor. One thing he won't be able to do, and that you should bear in mind before opting for a big wood burner, is to keep it fed and remove the ashes. Remember, the central heater is pumping warmth into every room in the house, and unless the place is really well-insulated, that takes a lot of wood. The only house I know well that was heated exclusively by a modern wood-burning central heater just happens to be the oldest place in town, and pretty drafty at that. The owners had to stoke it full at bedtime and get up good and early to add more wood to keep the water in the wash basin from freezing. And they celebrated reaching senior citizenship by installing an oil burner. However, manufacturers of more modern designs have reduced intervals between feedings to about 10 hours, or so they claim. Plus, the furnaces will accept really big logs, four-foot-cordwood, some of them. This greatly reduces cutting and stacking time too, needless to say.

The manufacturer and seller will be able to give you each model's heating capacity and fuel requirements. Just remember, in any kind of central heating, you are putting warmth into the entire house, not just a few selected warm spots as with stoves. More area, walls and floors are kept warm all the time, so you will need considerably more wood than with a conventional stove or two. I don't know anyone with a central-heating wood furnace that doesn't live on a good-sized country place, in the middle of a really large woodlot.

And then there are the ashes. The central heater will be in the cellar likely as not, and at season's end you will have 400 to 500 pounds of ashes on your hands. Well-combusted wood ash is light and fluffy stuff when fresh, and a pound takes up the better part of a cubic foot or more. Unless you wet it down in a lye tub you'll have to store or haul off a great deal of ash. Out on the farm where appearances didn't matter, we just threw excess ash up on the snow covering the garden. In town where that wouldn't be neighborly, the extra is stored in a brick ash pit that came with the house. Without it, we'd have to put the ash in sacks. If we had a central heater instead of the stoves, I imagine the cellar would be pretty well filled with ash sacks come spring.

Still, if you have plenty of wood and the ambition and equipment to get it into the woodshed, a wood-burning central-heating unit can get you as free of the oil companies as you wish to be—and in a house that's heated in the most genuinely modern way I can think of—an up-to-date central-heating system fired by the "old-fashioned" fuel of our petroleum-short future. And you can still have a wood range and all the fireplaces and heating stoves you want.

CHAPTER FOUR

Fireplaces, Old and New

About the same time that Benjamin Franklin was improving the heating efficiency of fireplaces by turning them into stoves, another Ben, one Benjamin Thompson, was calculating how to improve the fireplaces themselves. A remarkable man—like Franklin an inventor, diplomat and man of letters—Thompson unfortunately failed to side with Franklin and friends in the Revolutionary War and in 1776 was sent packing to England and out of American history. He did well for himself, though; earned the title of Count and took the name of his wife's hometown, Rumford—now Concord—New Hampshire. The improved firebox and flue design Count Rumford perfected is known as the Rumford fireplace, and his basic dimensions have never been improved for a good-heating brick-and-mortar fireplace. The illustration "Comparison of Fireplace Designs" shows and explains the difference in vertical proportions between a Rumford and a modern recreational fireplace. The great high mass of masonry in the Rumford absorbs and the angled back reflects most of the heat from the fire. Stand in front of one and you feel the gentle radiation six feet away. The combustion gasses cool by the time they reach the lintel—just warm enough to keep the flue drawing. The modern fireplace, on the other hand, is designed to waste fuel as efficiently as possible. The shallow, low firebox effectively puts most of the heat up the chimney. There's even an ash dump at the back so you can keep the firebox "clean" by shoving the hottest part of your fire down into the cellar.

Fireplace Basics

We touched on principles of fireplace operation earlier. Recall the American Indian tipi, where air for combustion and a

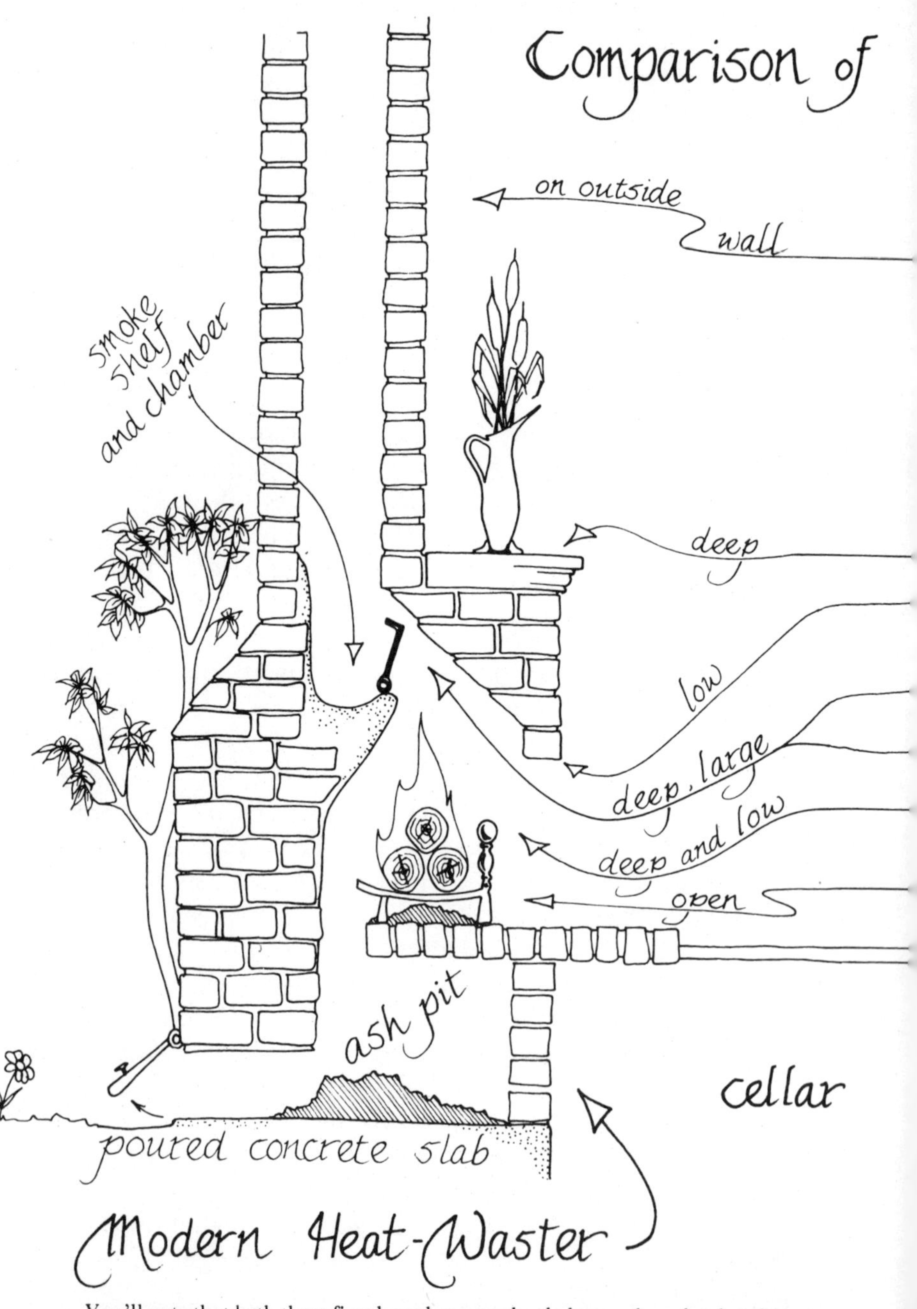

You'll note that both these fireplaces have smoke shelves and smoke chambers, so neither will smoke badly if proportions are correct. The big difference is in the size and shape of the firebox. In the modern design all the heat is sucked right out of the low little box and up the flue. In the Rumford, heat is absorbed by the masonry all the way from the solid ceramic base to the bricks in the breast up to the lintel. There is at least twice the heat-absorptive mass and more than twice the heat-radiating area in a Rumford than in a conventional fireplace.

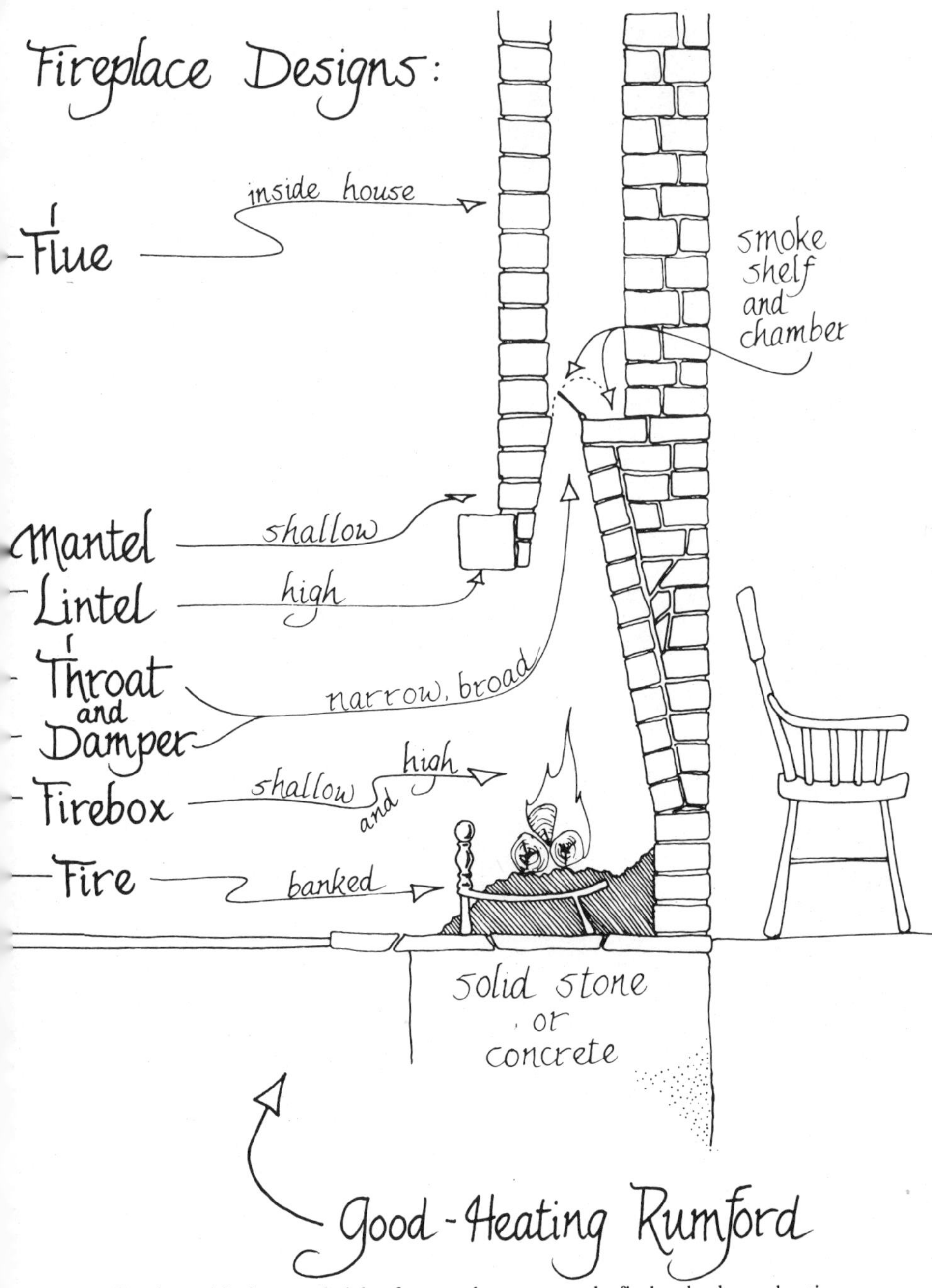

Further, with the great height of warmed masonry at the firebox back, combustion gasses will be drawn into the flue from well out in front of the lintel. And many users have their fire fronts a foot or so out into the room. That way a maximum amount of heat is radiated directly out into the living space. If you are going to build a new brick-and-mortar fireplace, make it a Rumford.

Here is an antique Rumford-design fireplace in the home of Howard Hastings of Barre, Massachusetts. There is no damper, and until one is installed, Hastings stuffs fiberglass insulation into the throat when the fireplace is not in use. Several of the fireplaces in this old home still do have the original "dampers," wooden screens that fit snugly into the firebox framing to keep in warmth when the fire was out.

healthy atmosphere came in through small openings at the bottom of the tent. Draft openings in effect. And the top flap was regulated so that air flowed in at the top and circulated down to be warmed, then took the smoke out as it rose. This is how an efficient fireplace operates, or how it should. Ideally, only enough air enters the flue from the room to remove combustion gasses from the firebox. David Howard, an architect and builder from Alstead, New Hampshire who puts Rumford fireplaces in his post-and-beam houses, has a simple test of a fireplace's drawing power. If the smoke from his pipe, released at the lower edge of the mantle, is drawn gently, slowly but completely into the flue, the draft is right.

If the firebox lacks a smoke chamber and empties right into the flue, as is the case in many older fireplaces and, unfortunately in some new prefabricated designs, you will probably

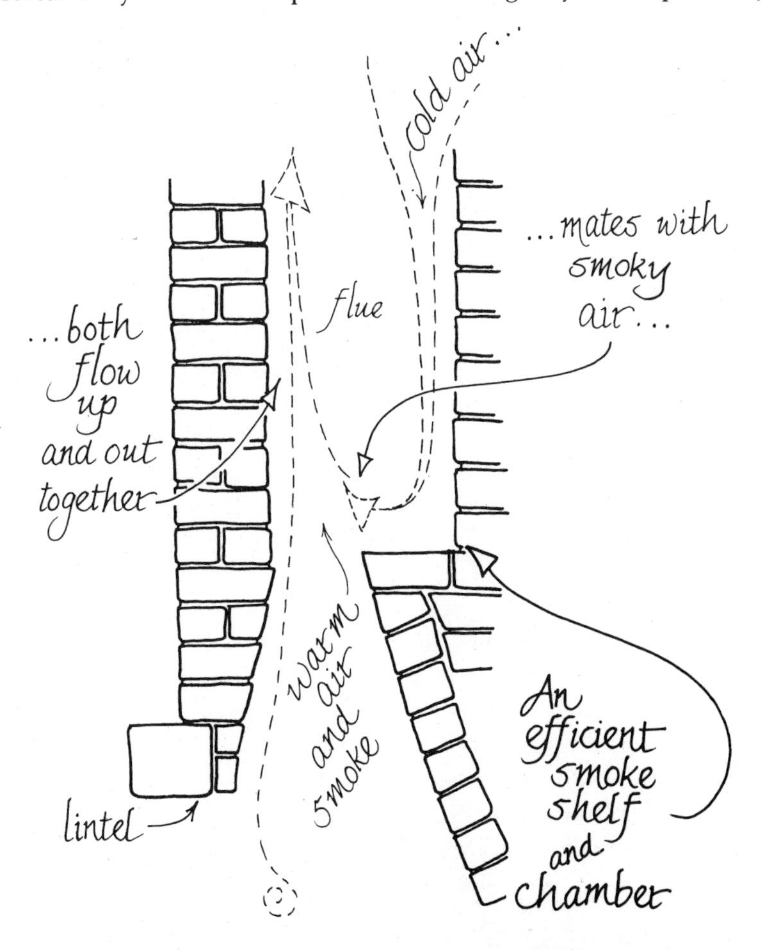

have a smoky fireplace. That cold air that pours down the flue will come right into the room, carrying smoke with it. So, in Rumford and other efficient designs the smoke chamber has a pronounced shelf at the bottom just above the firebox. Outside air flows down, hits the warm shelf and mates with the smoky air entering from the firebox, then all flows happily on out.

For proper functioning, you must maintain strict ratios between size of fireplace opening, size and height of flue and so on. The figures are given in the following pages; deviate from them and you are asking for trouble. To explain what problems you might bring on yourself let's examine the reasons an existing fireplace might not perform properly and suggest solutions. Use any ideas that fit problems you may have with your own fireplaces.

Inspecting the Facility

Before lighting a first fire, check the fireplace for bad joints, cracked bricks and all, just as we looked the flue over earlier. The inside of the firebox may be of brick or soapstone or other type of rock slabs if it is an older one. Circa 1920 fireboxes are often lined with large fireclay blocks and in newer homes you'll usually find a steel liner. Check joints in all ceramic boxes and patch any small holes with furnace cement. Remove any loose bricks and replace with fireclay mortar.

If you've a metal liner, pound around looking for holes. You may be able to patch a small hole with furnace cement or put in a bolt or screw to seal it up. If there are big holes or if the damper mechanism is rusted through, call a mason. The problem with the metal liners is that they do deteriorate in time, particularly the less-well-made ones, and repair is difficult. You may have to have a new firebox put in. Make it a brick-and-mortar Rumford—coming up a bit farther on.

With all fireplaces, get your hand as far up into the flue or smoke chamber as possible. Remove any fallen bricks, birds nests or whatever. Here is one advantage of modern design; the big opening gives better access to the smoke shelf than the narrow Rumford throat.

Hopefully, there will be a fireplace damper in the throat, a metal door that closes the flue when the fireplace is not in use. Be sure it opens and closes tightly. It should be hinged at the back

and the lever, handle or other control should open it so the metal plate rests fully vertical. If there is no damper and you live in a chilly climate, call the mason and have one installed. Until it is in, stuff rags into the throat when the fire is cold, else all your warmth will go up the flue.

Lighting Up

You should have a set of andirons or log rests; if not, get a pair that extends the full depth of the firebox. If you have a fire grate or box arrangement it was most likely designed to burn

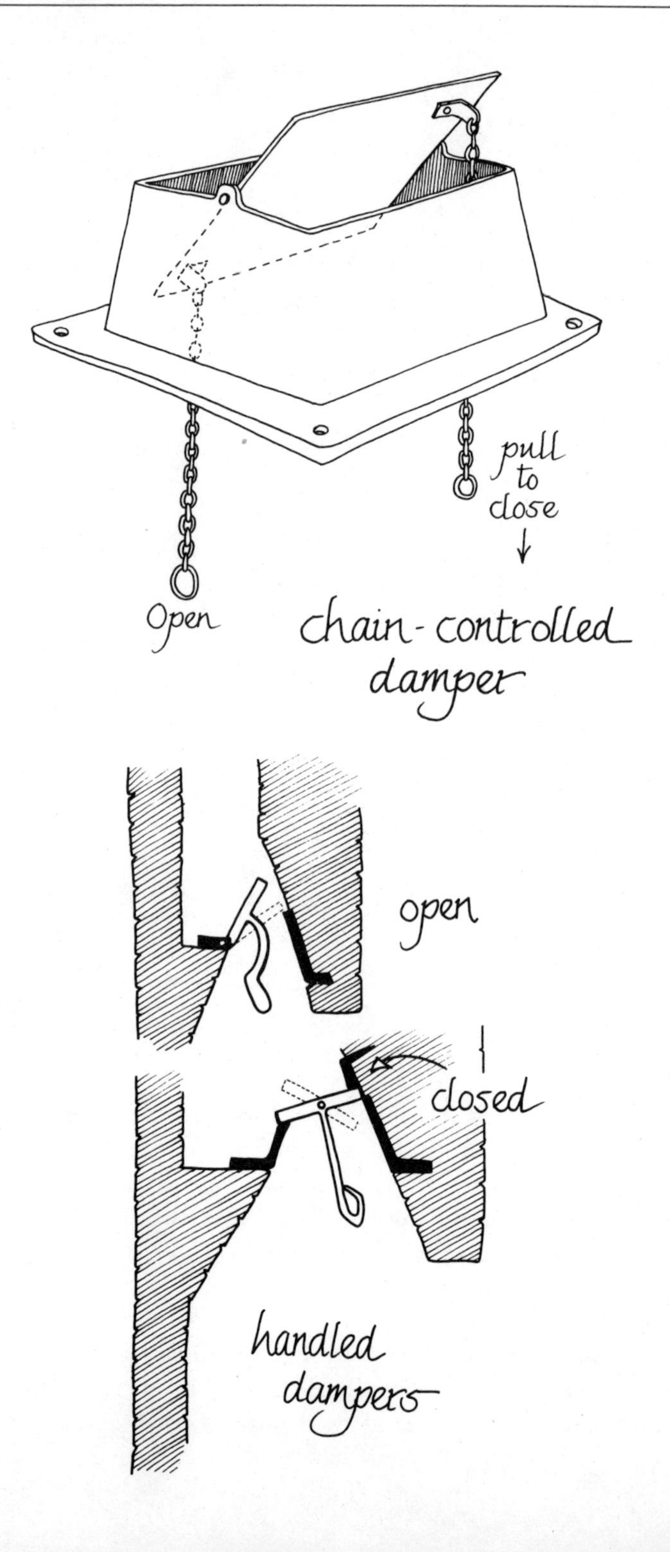
pull to close
Open
chain-controlled damper
open
closed
handled dampers

coal and will not hold a properly banked wood fire. Let's build one. Open the damper all the way and wait a few minutes, particularly if the weather is cool and you have a typical modern chimney, built on the outside of the house. Cold chimneys don't draw, remember.

And remember to let the wood warm too. If you try to start a fire with frozen logs just brought in from outside you'll have a long wait. The kindling in particular should be room temperature. Crumple up several sheets of newspaper—a *loose* crumple—and put them on the firebox floor. Take a good-sized, preferably *unsplit* log, lay it on the andirons, and push it all the way to the back of the firebox. Put another big one, preferably *split* out in front and fill the space between them with kindling sticks laid so plenty of air can get through them. Then on top of

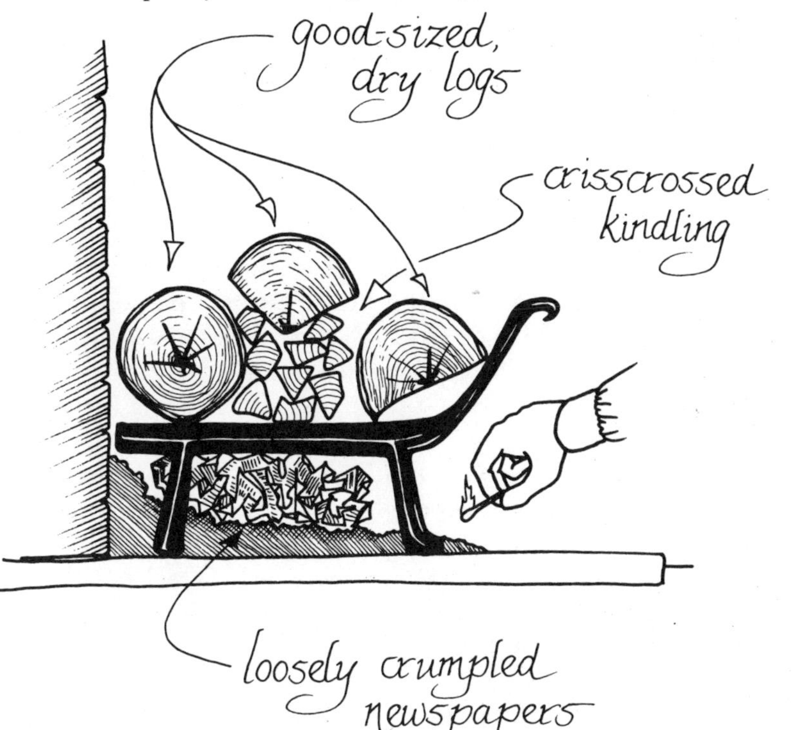

Laying a proper Fire

the kindling lay another split log. Arrange the three large logs so they are almost, but not quite touching along most of their lengths, in Louise's magic wood spacing. Just as it did in the stove fires, a narrow opening between logs increases velocity of the draft flowing up from underneath, in effect "blowing" on the fire to increase oxygen supply. Not an effect you want in a good heating fire, but it's necessary to get the blaze going.

Now, light the paper. If you've a good draft and the wood is properly seasoned, split the right size, and warm, you should never have to use more than two sheets of newsprint to start a fire—if you've got the knack, that is. I can still go through half the Sunday paper in front of a fireplace at times.

Keeping the Fire Burning

Once the fire is going, your object is to create a heavy bed of ashes; indeed the entire floor of the firebox, at least up to the top of the andirons, should be solid with warm ashes. The heart of the fire is a body of red-hot coals in the center of the bank that is slowly settling down as it goes to ash while radiating out heat with maximum efficiency. Indeed, once you've the fire established, you should never see a flame more than a few inches high. Logs go on one or two at a time, and are continually pushed back into the ash bank. Remove excess ashes as they are pushed out onto the cool front of the bank and to the sides. This way you can light up once in the fall and keep the same fire till spring. Of course, it will take some time to build up the first ash bank and if you let it get cold, you'll have to start over each time with a new fire. However, if you're seriously heating with wood, the well-banked, smokeless and nearly flameless permanent fire is the kind to have.

Incidentally, there exists considerable difference of opinion around the country on just what a well-banked fire consists of—indeed what the term "banked" means. According to the dictionary it means to cover the fire with fresh fuel and reduce your draft to put the fire in an inactive state. Well, hereabouts, the bank of ash-covered coals is part of a well-banked fire. You build up that ash bank, use the ashes to build a bank (the word also means a mound as of earth or ash) over and around the coals, then bank up the ash bank with another bank of logs. If you closed the damper down and told it to remain inactive as the

dictionary says, your house would be smoky pretty quick. Fires can't read. Or hear. But they sure can heat if you build them right, banked three ways, I guess. (That dictionary definition must be a semimodern reading of an old term. You could do as it says in a coal furnace, a foundry or in the engine of the old Wabash Cannonball. But not in a fireplace, or any stove that I've ever heard of.) The banked fire is vastly more efficient than any ornamental blaze (though there's no harm in tossing on a few small logs for a cheerful flame in the evening). But, in the ash bank your logs will turn to coals from the bottom up, burning completely but slowly so that the maximum amount of the heat stays in your room and none is lost in those cheerful, but wasteful high flames.

The Smoky Fireplace

Now, what if your fireplace smokes?

The most common cause of smoking in newer houses is a

lack of air for the updraft. Turn off the kitchen or bathroom exhaust fans and open a window in the next room a crack. If the resulting draft through the room and along the floor is a nuisance you might want to drill holes in the wall on both sides of the fireplace. There are small, closable registers for such use. If smoke comes in puffs, but never continuously, you've a chimney problem, most likely an inability to handle strong wind gusts. Either install a wind barrier (see page 41), or raise the flue (see page 169). A minor, but naggingly constant bit of smoke in the air can often be eliminated with a smoke curtain or hood that effectively drops the lintel. A lintel that is too short is a feature of many a poorly designed fireplace. An old, but dangerous practice is to hang a thick cloth from the lintel. A metal strip, hood shaped or simply flat, can be mortared in place — and won't burn. Perhaps after experimenting with a strictly temporary "fire or smoke curtain" you might want to put in another lintel support and drop the masonry several inches.

If you are still choking on smoke put out the fire and measure all dimensions of the fireplace, flue and smoke chamber and compare them to the figures in the chart. Most commonly you'll find the throat too small or too large and situated too far

Dimensions for conventional modern fireplaces and flue sizes.

(Letters at heads of columns refer to drawing below)

Size of fireplace opening						Size of flue lining required	
Width	Height	Depth	Minimum width of back wall	Height of vertical back wall	Height of inclined back wall	Standard rectangular (outside dimensions)	Standard round (inside diameter)
w	*h*	*d*	*c*	*a*	*b*		
Inches	*Inches*	*Inches*	*Inches*	*Inches*	*Inches*	*Inches*	*Inches*
24	24	16-18	14	14	16	8 x 13	10
28	24	16-18	14	14	16	8 x 13	10
30	28-30	16-18	16	14	18	8 x 13	10
36	28-30	16-18	22	14	18	8 x 13	12
42	28-32	16-18	28	14	18	13 x 13	12
48	32	18-20	32	14	24	13 x 13	15
54	36	18-20	36	14	28	13 x 18	15
60	36	18-20	44	14	28	13 x 18	15
54	40	20-22	36	17	29	13 x 18	15
60	40	20-22	42	17	30	18 x 18	18
66	40	20-22	44	17	30	18 x 18	18
72	40	22-28	51	17	30	18 x 18	18

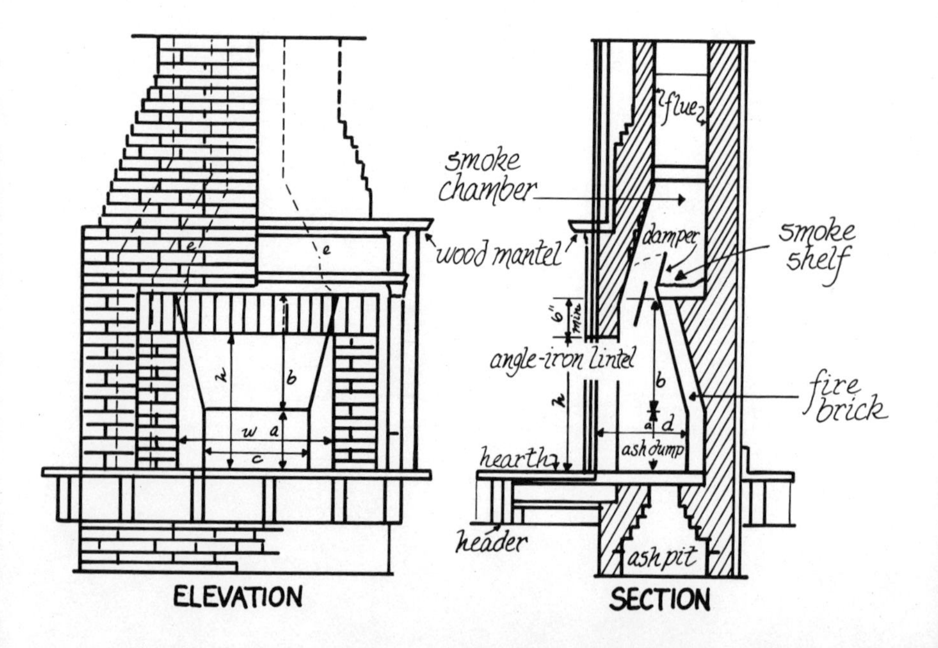

back in the firebox. It should be at the front and its area (length X width) should be at least equal to the flue area. With a too-big throat and an adjustable damper you may be able to choke the throat down enough to improve the draft. If the firebox opening is too large for the throat/flue relationship, you can modify the dimensions with firebrick at the back, sides or on the bottom. Before actually mortaring in any changes, I'd experiment by just laying up regular building brick and trying out the new dimensions with a small, but smoky fire. Often simply raising the hearth can help a lot. That is what we did back on the Pennsylvania farm, where the fireplace had been built without any smoke chamber at all. Raised a foot, and with a foot-wide black iron curtain to drop the lintel, the smokiness just about disappeared. To have attempted to reconstruct the fireplace and put in a proper smoke chamber was just too much of an undertaking for our limited know-how at the time, and I'm still not sure of how to go about such things as supporting the flue top while reworking the bottom. Had we done it, a professional mason would have been called in, and I'd advise you to do likewise if relatively simple measures fail to cure your smoky fireplace.

One other corrective measure we have learned about since our smoky fireplace experience is to lower the lintel if needed and mortar in a sheet-steel angle at the back of the fireplace, in effect making your own smoke chamber. This would probably not be possible in a small fireplace and I don't know but that the esthetics would prevail over a bit of smoke in the air with an antique walk-in. But the drawing shows what I'm getting at and it may work for you.

A common problem in old-but-not-antique flues are cracks or open joints in the "wythe," the wall separating your fireplace flue from a central-heating flue. Air passing between flues will interfere with draft in one or both and the traveller and cement treatment described earlier will help. Other flue problems can include too steep angles, bumpy flue linings and improper height. There's not much you can do about any but the latter problem.

Recall, the chimney should rise at least two feet higher than the roof peak or any tall, nearby objects. Add on another foot of height if your roof is flat or nearly so. Also, measure the area of the fireplace opening and height of the flue from the firebox bottom. The opening should be no more than 12 times the area of the flue's inside dimension with flues over 22 feet high. If the flue is less than 22 feet high, it should be at least one-tenth the area of the fireplace opening. If the chimney is too low, build it up with bricks or whatever else it's made of or tack on a ceramic smoke pot or steel rotating ventilator. A sheet metal tube just rammed down into the flue can help if you don't care much about appearances. For details, check back through the chapter on old flues.

Improving the Efficiency of Existing Fireplaces

Now that you've made sure your fireplace doesn't smoke overmuch, let's see what can be done to make it a better heater. As it is, I'd bet you are losing 90 percent of your heat up the flue.

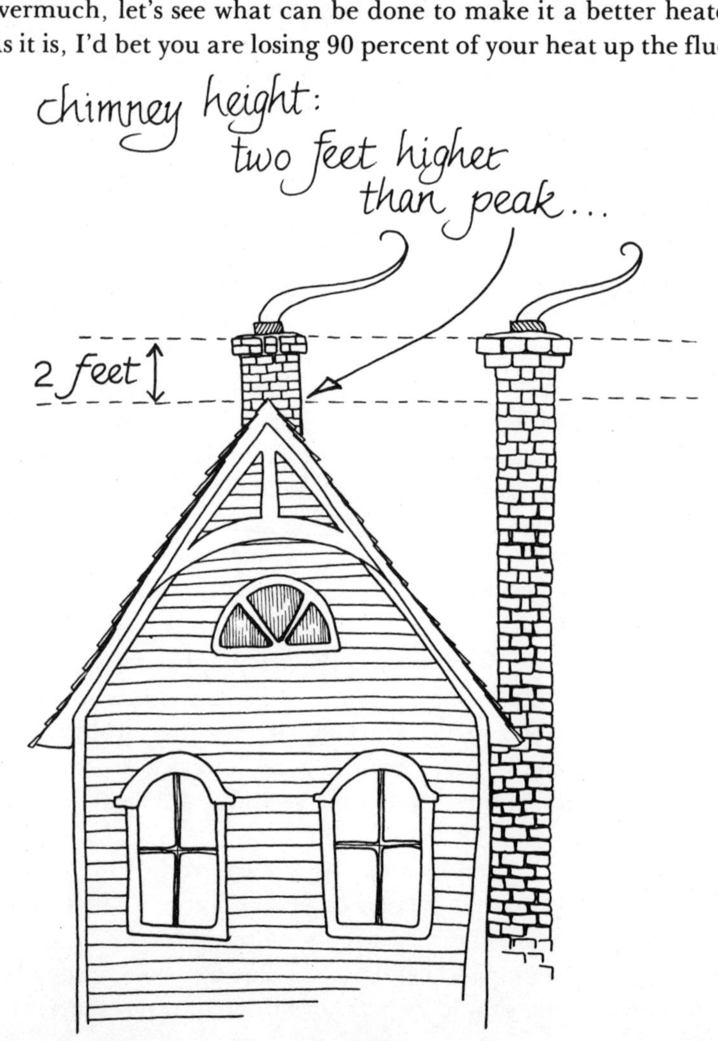

One thing is to learn to use the damper. Most commonly, dampers come with a notched handle or other device for regulating the size of the throat opening. Get that well-banked fire going as far out in front of the firebox as you can and see how far down you can close the damper. If your damper only has two settings, open or closed, see if there isn't something you can do to give a series of intermediate settings by tightening bolts or filing notches in the handle. Experiment by stuffing asbestos board or brick into the opening to see if restricting the throat size is practical. You may even want to replace the damper with an adjustable model.

There are a number of patented grates and other devices on the market now for getting more heat out of the fire. Garden Way Research, Charlotte, VT 05445 sells one originally designed to raise the fire and cure a smoky fireplace. But as the illustration shows, the firegrate is higher in front than in back. Coals fall down, radiating heat out into the room—sort of raising the fire's skirts a bit.

Another variation on the theme is made by the Dahlquists of 31 Morgan Park, Clinton, CT 06413. Their Radiant Grate raises her skirts even more and they sell reflector ovens, a stove and a grill to cook by radiant heat. All their meals are cooked before the fire. And from the photos I've seen, their fireplace is a conventional fuel-waster—*without* the cooking value provided by a grate, that is.

You've probably seen ads for tubular grates, a series of **C**-shaped tubes welded on legs to face out into the room. The idea is that cool air is drawn in at the bottom, rises as it is heated by the fire, then flows out to warm the room. Some come equipped with blowers to increase the air flow. If you don't mind a blast of hot air in your face each time you throw a log on the fire, they work. But, beware if you buy one. Many are coming out made of nothing more than auto exhaust pipe painted black to resemble cast iron. I don't know of any that come in real cast iron, but shop around for the heaviest gauge pipe you can find. The thin stuff won't last through two heating seasons. And I'd think that the ones with a blower would last longest as the greater air circulation would keep the pipe cooler than natural air flow.

Garden Way's Grate

As wood heating increases in popularity, more and more new ideas will be coming onto the market. I've seen ads for other devices that run air through the top or bottom of the firebox—big metal tubes formed in a C-shape, or serpentine arrangements with a fan at one end and a lot of hot air at the other. These, at least improve on the grates in a series of vertical C's, which effectively blow the hot air right up to the ceiling.

Any of the patent grates will run $50, more or less, and the electricity the blower-equipped models use is negligible compared with the Btu's of heat they keep from being wasted up the flue. I'd say that any well-made one is a good long-term investment if you are stuck with a conventional fireplace. Still, none of them will make a heat-waster into a Rumford.

Most fireplace accessory stores sell fire screens made of wire mesh and others of glass. I don't like the mesh between me and the heat, but I wouldn't leave the room for long with a fire going without one. (And neither of us would leave the house with a live sparking fire going under any circumstances. A well-screened ash and coal bank with virtually no new wood on top—a solid ash covering, but enough coals beneath to hold heat for hours—is safer.) But those glass doors are calculated to turn a normal

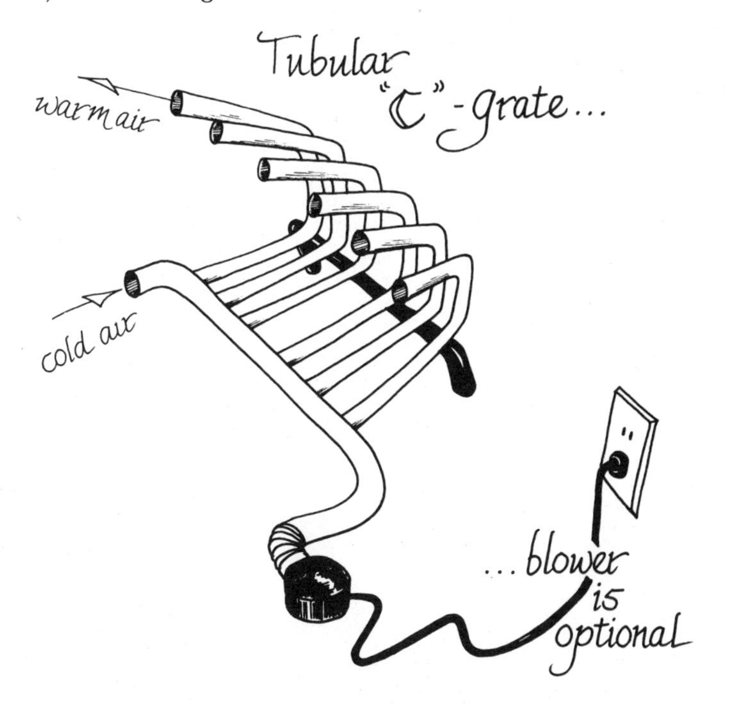

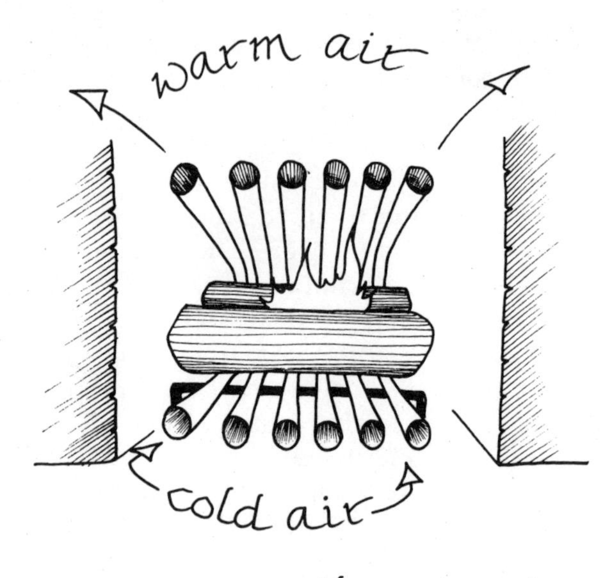

... another model

wood-wasting fireplace into a super squanderer. The glass keeps in virtually all the heat. Draft is brought into the fire through slots at top and/or bottom, though, which is reminiscent of wood stoves.

What if a glass screen was fitted with adjustable air controls and a damper was installed so as to be adjustable from outside the fireplace, and if the entire firebox was pulled inside the house, and if it was as thin as safe and practical? Sounds like a description of a stove. You may as well put a stove in.

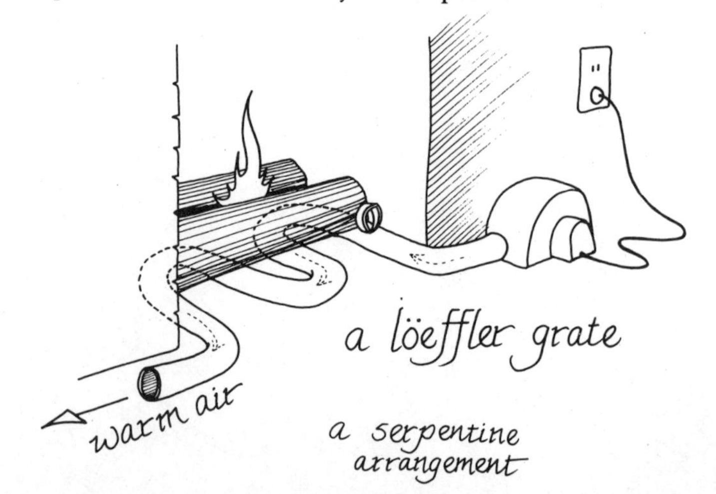

a löeffler grate

Building a Fireplace

Fireplace construction is about the ultimate skill in the mason's trade, and trying to explain it in detail is beyond the scope of this book. It's also beyond the scope of this writer's skills and first-hand masonry experience. Stone walls and flue maintenance and such simple stuff as mud-and-stick or precast cement-block flues I've done. But take a hard look at a brick-and-mortar fireplace. The inner surface of the firebox, smoke chamber and flue should be perfectly flat and smooth so smoke will flow freely. And look at the angles of those back firebox walls. The back slopes up, out and toward you. The two sides do the same thing but at a biased angle. It takes a great deal of time and the advice of an experienced mason. Bricks have to be beveled at angles in the box and woven where walls meet, and everything has to end up plumb, square and smooth. An amateur *can* do it, but not by simply reading a book.

What attempts I have seen in books to describe fireplace construction are nothing but paraphrases of a government publication—and *it* tells you on page 1 to get an experienced mason. The booklet is *Fireplaces and Chimneys* (Farmers Bulletin No. 1889), issued by the U.S. Department of Agriculture (Send 20 cents to the Superintendent of Documents, Washington, DC 20402.) You will also find a selection of books on the trowel trades in most public libraries.

However, no books I've seen adequately describe the dimensions and structure of a good-heating Rumford-style fireplace. You can get the Count's journals which ramble on about it in eighteenth century English in most public libraries (Harvard House is the publisher). Or you can send $2.50 to Yankee, Inc., Dublin, NH for *The Forgotten Art of Building a Good Fireplace,* by Vrest Orton. The book contains a lot of philosophizing that some folks might think more appropriate to Count Rumford's time than our own, though Rumford's basic fireplace dimension principles are there too.

Rumford insisted on characteristics that modern builders have found unnecessary—such as requiring that a plumb line dropped from the center of that throat lands squarely at the center of the firebox. Presumably that perpendicular should go straight up the flue, so smoke would also go straight up without

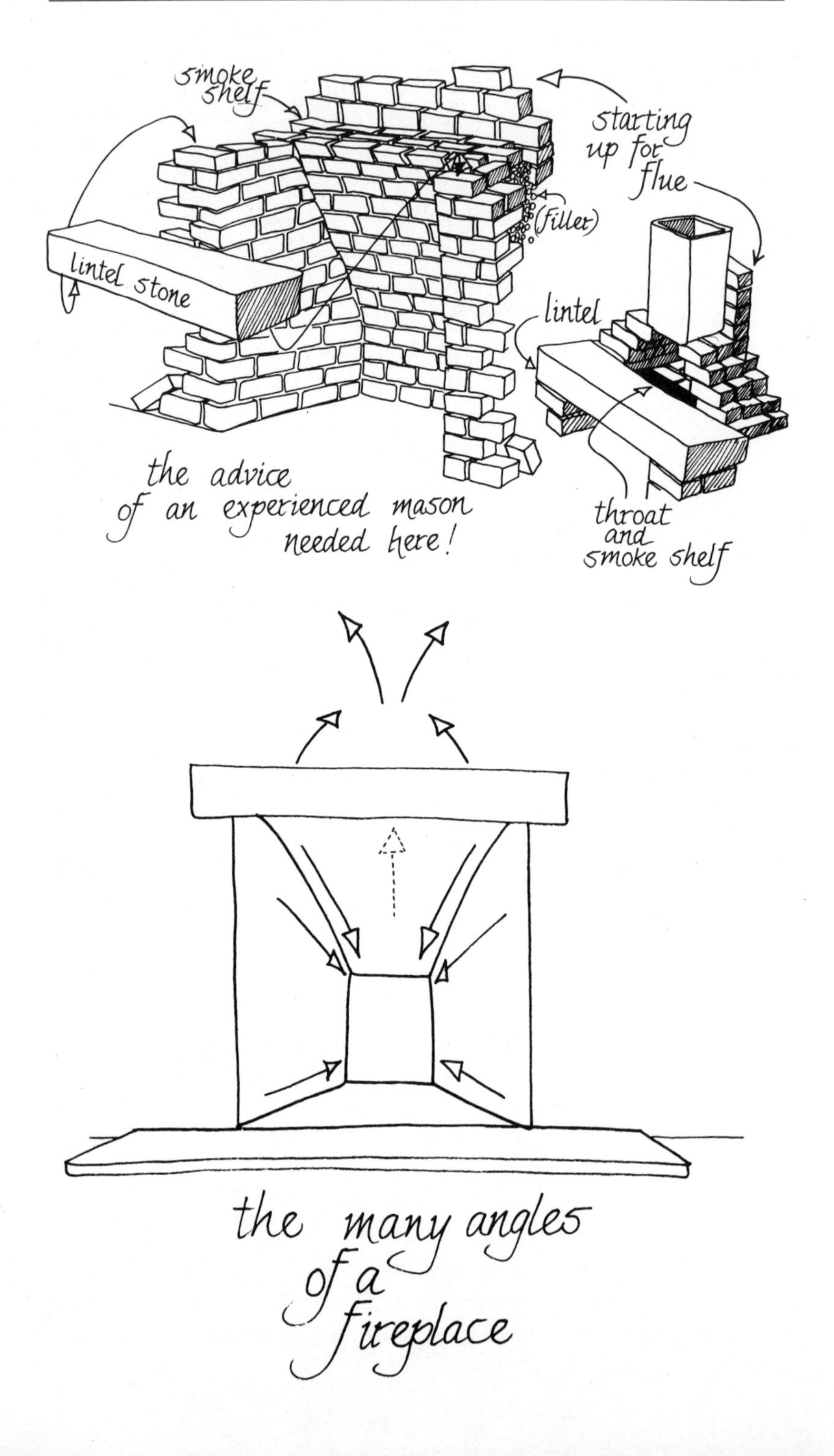
smoke shelf
starting up for flue
(filler)
lintel stone
lintel
the advice of an experienced mason needed here!
throat and smoke shelf
the many angles of a fireplace

This is an example of colonial restoration and fireplace construction at its very best. Mason Jim Dowd is constructing a late 1700s kitchen fireplace, complete with beehive oven.

hindrance. I wonder if that rule was Rumford's way of getting people to build in a minimum of angles in a flue; by his rule you could look up the throat and see the sky which is the case in some of the best-drawing antique fireplaces I've seen. However, modern masons know you can put in as much as a 45 degree angle if absolutely necessary though chimney height may have to be increased to compensate. Better is to keep angles down to 30 degrees or less.

Planning for a Good-Heating Fireplace

Though I'll not attempt to tell you or your mason how to build a fireplace, I will offer the basic principles that Louise and I considered when trying to decide between a fireplace and a stove to heat our present house. Perhaps you'll find them useful to your own planning for a new house (build the house around the fireplace) or installing one in an existing dwelling.

First, the fireplace and flue are the heaviest part of any home and if you go with ceramics from cellar to roof peak, the structure needs its own foundation. An eight-inch-thick, poured concrete slab is usual in modern construction (where tract houses are built to last only 15 years). If yours is an older house, like ours, built to last, or if you are building one to last, you would want a solid pier of cemented rock, concrete block or poured concrete. Make it six inches larger than the base of the fireplace and any other flues. Have it extend from well below frost line through the cellar to the first floor. In any existing house that means sawing a big chunk out of your floor beams and surfaces.

Next, if you are serious about heating with the fireplace, you will want the entire brickwork within the house spaced so heat will radiate out from all sides. As I've said before, putting the fireplace and flue on an outside wall is easier to build, especially in an existing structure and we put one in ourselves. But for heating, it is wasteful, and if you can keep chimneys inside, you should do so.

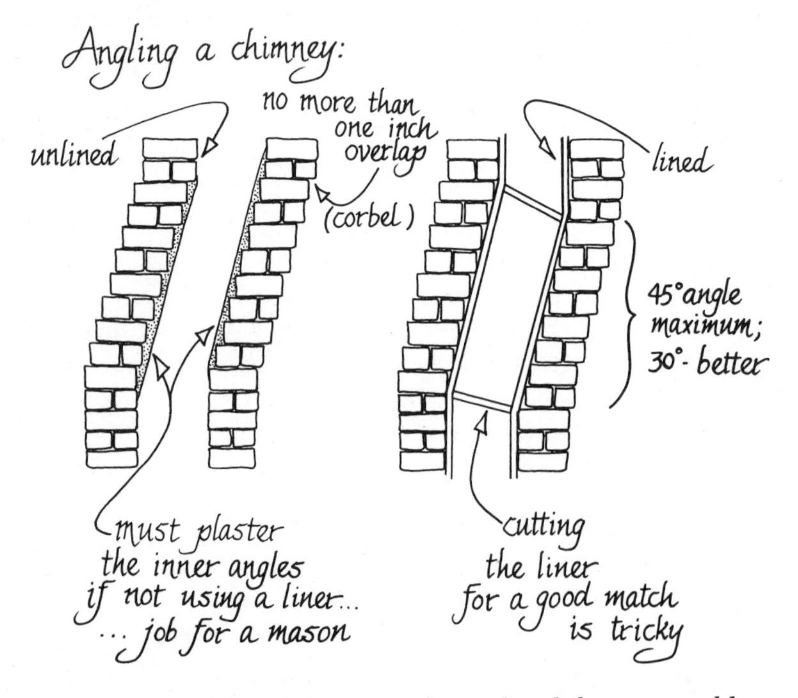

I mentioned the 30 degree optimum bend that you could put in if needed to get the flue around plumbing or structural items that you can't or don't want to move. Another rule is that the ceramic must never be closer than two inches from any flammable material, and that includes all beams and rafters. Then there is the problem of getting the chimney works through the upper floors and the roof, each level requiring a major rebuilding job. As it turned out for us, at the new flue location that made best heating sense in our house (the center alongside the existing flue) we would have been forced to cut through all main supporting timbers from cellar to roof. Too much.

What we did consider, briefly, was a compromise all around: a good-heating ceramic structure from floor to ceiling in the living room, and prefabricated stovepipe on up through the upper stories and the roof. Recalling the old plank-supported fireplace on our first farm and after talking with a house mover, we planned to support the mass as follows: sister beams along existing joists, six-inch-thick oak planks bedded in a leveling layer of mortar under these, held up by ten-ton house jacks (six, probably) set atop a concrete-block pier. Or at least that was the only way I could figure how to do it myself without having cement mixers chewing up the lawn.

The only problem we didn't get solved to our satisfaction, and a question that surely would have taken the advice of a builder and probably the fire marshall, was how to prevent the flooring from catching fire. Perhaps a high hearth set on perforated brick (building brick with two or three holes in it) or porous cinder block would have done it. I considered laying perforated brick with holes aligned all the way through and fixing a blower to one side, keeping the brick cool and dispersing heat (perhaps through piping to the upper floors). Though taking out floor beams would have been too much of a job, we could have easily torn up the flooring and replaced it with a noncombustible, nonconductive material, most likely fine quartz sand.

Let me reiterate: we have not yet done this and don't know if it would work or satisfy our building codes, which are in part set by and to make more work for the housing industry. So, I make no recommendation for it. However, everyone these days is coming out with wild ideas on how to get the most out of wood heat, so I thought I'd offer one of my own. If you want to try it, I should pass on the evaluation of master mason Jim Dowd, a Rumford builder from Hardwick, Massachusetts: "Too expensive, impractical and won't last." A man of few words, Jim is rebuilding an old colonial fireplace that rests on rotted timbers, and he oughta know.

Designing the Fireplace

Back to Mr. Thompson, or Count Rumford, and his fireplace. Assuming you have your foundation set, mason with trowel in hand, what's next? First, decide the general dimensions of the finished fireplace, hearth and trim. Interior design is

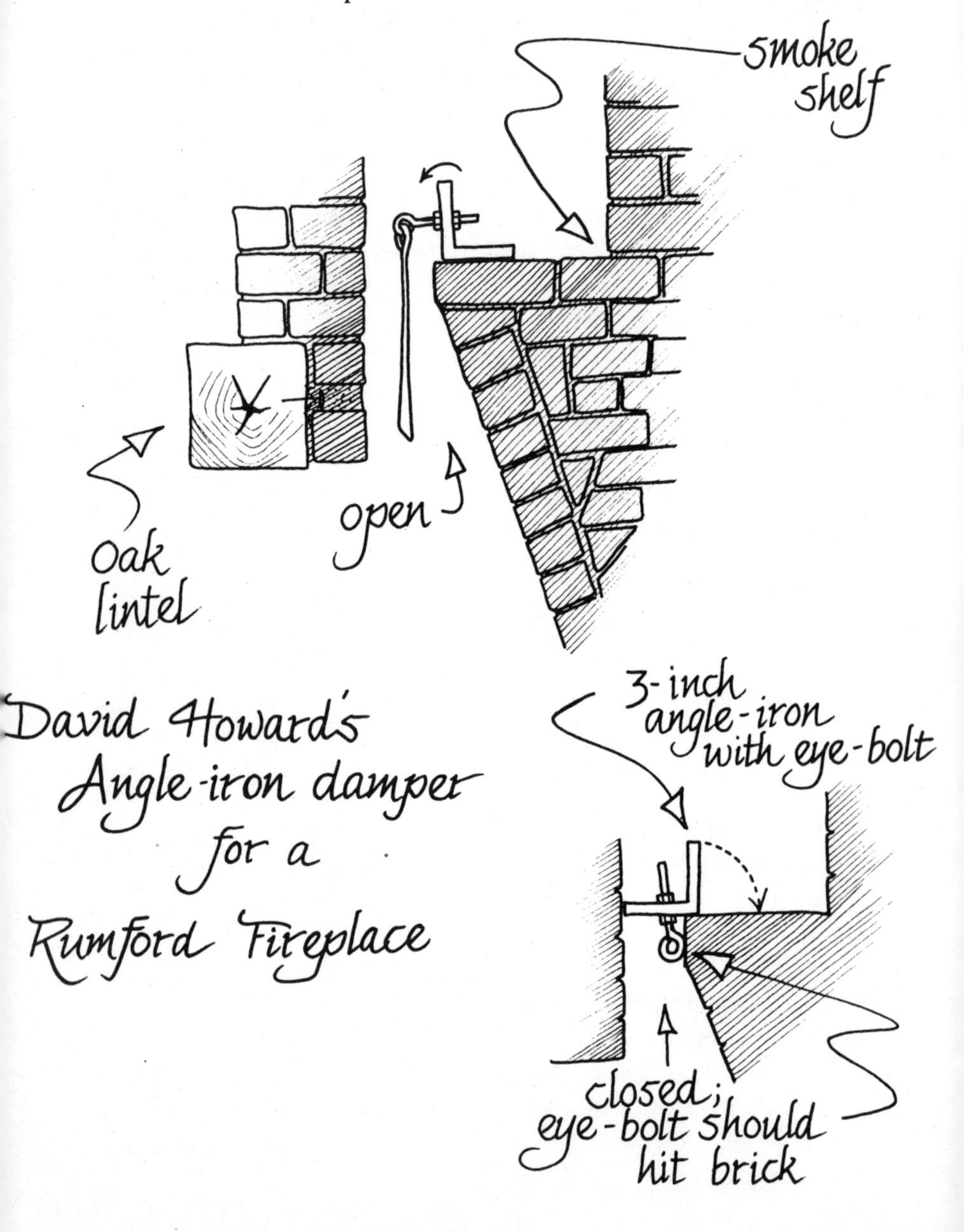

David Howard's Angle-iron damper for a Rumford Fireplace

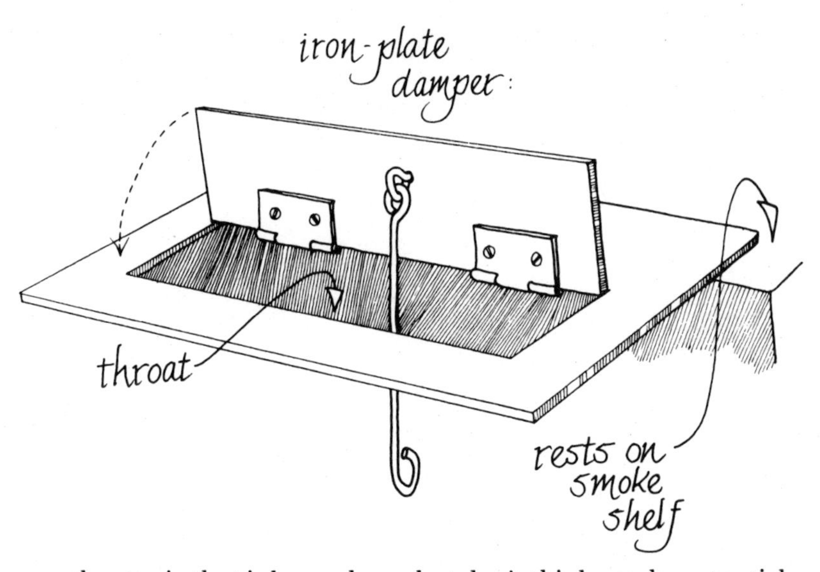

another topic that is beyond me, but don't think you have to stick to Early American decor just because the fireplace is based on old principles. It will look fine in any setting. There are several books out on fireplace design. All contain more or less the same technical information you'll find in the USDA manual. None I've seen will help you heat with wood, but they do have pictures of all styles of rooms containing handsome fireplaces cheerily wasting wood.

We've combined Count Rumford's principles with the practical know-how of several men who are building modern versions of his fireplaces to compare dimensions of several Rumford-style fireboxes with a modern fuel-waster. The differences are not great, but they are crucial: the firebox shallower, higher and angled forward and lower down, the throat more narrow and extending all across the front of the firebox. Your mason will have his own ideas about size and shape of the smoke chamber, what bricks to use where and so on. Just make sure he sticks to a basic Rumford firebox. Oh yes, you'll find no damper made for a Rumford. David Howard (whose 40-by-40-inch design is included in the chart) uses nothing but a length of three-inch angle iron with a rod attached. Just flip it up or down for open or closed. And you can unhook your operating rod, turn the iron over and slip it down and out to get at the smoke shelf. We've included a sketch of a more conventional damper you can have made at any job shop. Good building.

COMPARATIVE DIMENSIONS OF FIVE FIREPLACES

(in inches)

Type of Fireplace	Firebox Width	Firebox Height	Firebox Depth	Fire back FW.	Fire back FH.	L.	Throat TW.	Throat TD.	Flue
Rumfords									
Small Upstairs	24	24	8 to 12's	8 to 12's	8 to 12's	8	24	3	8 x 9
Yard-wide	36	36	12	12	15	15	36	3	10 x 13
Cordwood	50	50	17	17	20	12	50	4	13 x 16
Howard									
Yard-wide	40	40	18	17	17	12	40	3	12 x 16
Modern									
Conventional Yard-wide	36	29	17	22	14	12	*	*	8½ x 13

*The throat dimensions of conventional fireplaces vary, but invariably the throats are too big or too small.

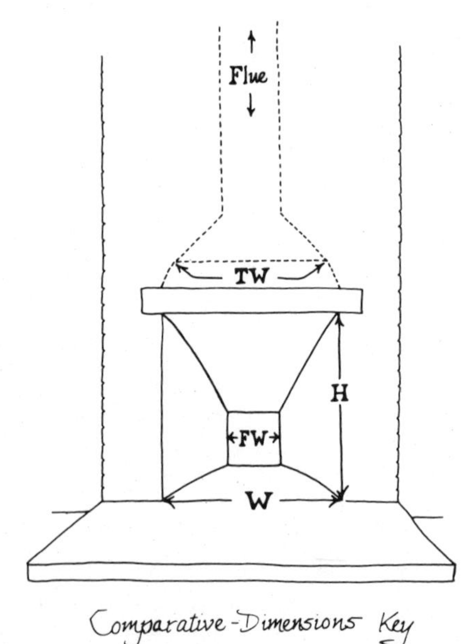

Comparative-Dimensions Key

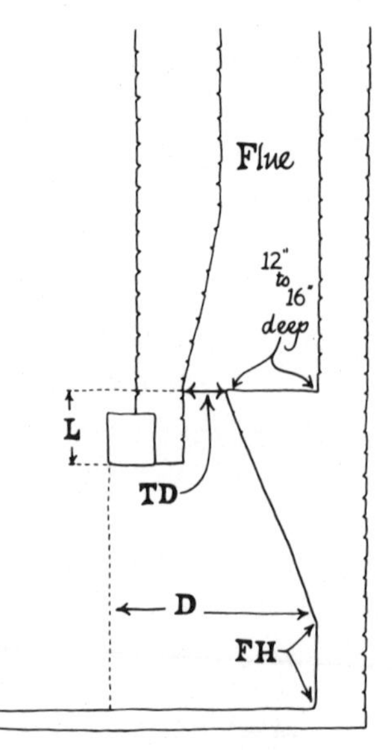

To Build a Rumford Fireplace
Basic differences from conventional modern fireplaces

1. *The throat is as wide as the firebox, but only from three to four inches in depth. (Some say to four-and-a-half inches.)*
2. *The width of the front opening is two to three times the depth of the firebox (usually three and always equal to the height).*
3. *The height of the front opening is two to three times the depth of the firebox (usually three and always equal to the width).*
4. *Height of fireback, where it begins to slope forward is usually 12 inches, less in small boxes.*
5. *Width of fireback is equal to the depth of the firebox.*
6. *Smoke shelf is as wide as the fireplace opening and 12 to 16 inches deep. (Some experts say 6 to 12 inches deep.)*
7. *Smallest permissible flue size is the larger of: one-tenth or twelfth the area of the fireplace opening depending on flue height—or—the area of the throat.*

Prefabricated Fireplaces

As mentioned, a brick-and-mortar fireplace just wasn't practical for our place so we halfheartedly looked over the prefab fireplaces. These are steel and come in two styles. One is free-standing, most typically shaped like a big tin funnel over a base shaped like a giant tuna fish can. The can is so well insulated you can set it in the center of your room as is, hang the funnel from the rafters, run an insulated flue up through the roof and have a fire that evening. Some of them come in bright colors that match the modern (mod?) decors, and I hear they are big in Southern California. There they dig a firepit in the stone floor, surround it with a tier of seats, hang the funnel from a really big and powerful flue and toast their weenies. Of course the things don't heat for a hoot, but who needs heat in Southern California?

The other type of prefab we looked at comes as a conventionally designed fireplace, complete from base to top of the

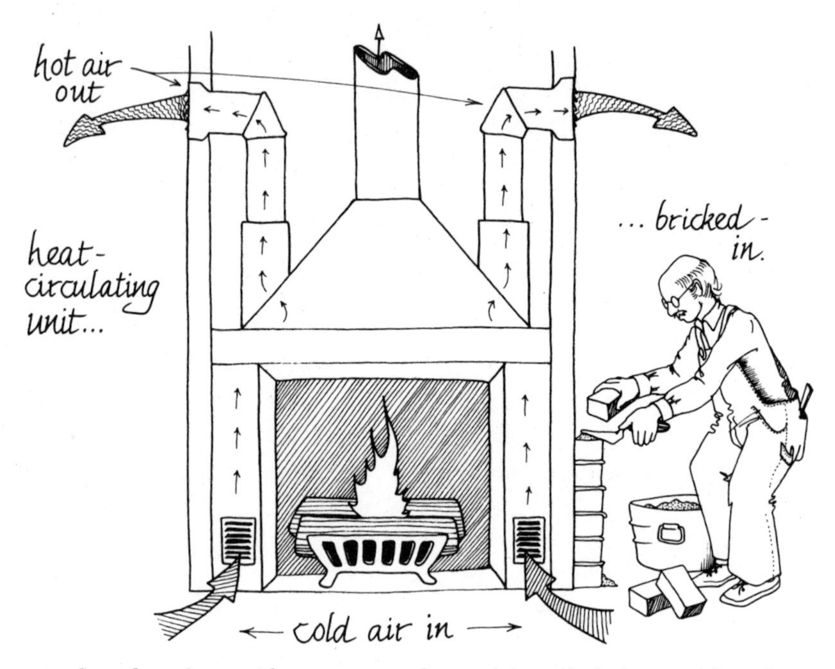

smoke chamber. They are steel, multi-walled for perfect insulation and can be placed right on a wood floor, boxed in with two-by-fours resting right against the outside and covered with wallboard. Then you paste on plastic brick, connect your prefab flue and sit back to look like something out of the Sears, Roebuck catalog. If I'm being overly snide about these things, I apologize. They are cheap, quick to go up, and don't at all have to be faced with plastic, but can be bricked in or faced with stone or lovely colonial woodwork. They will look just like the real thing. But that's the problem. They are all for looks. Especially the superinsulated jobs. They are *so well* insulated they are the ultimate in heat-wasters. They suck all the wood heat, plus your room heat, up the flue without even providing the consolation of a warmth-radiating brick firebox once the fire is banked and damper closed a bit for the night.

Air-Circulators

The only prefabricated fireplaces that have any right to claim even a minor heating capability are the air-circulators. Again, two kinds. One is a standard steel firebox. You build a box, brick usually, around it. On each side, perhaps top and bottom if it is built into a flat wall, you install ventilators. Air flows in at the bottom, is heated by the firebox walls and comes out at

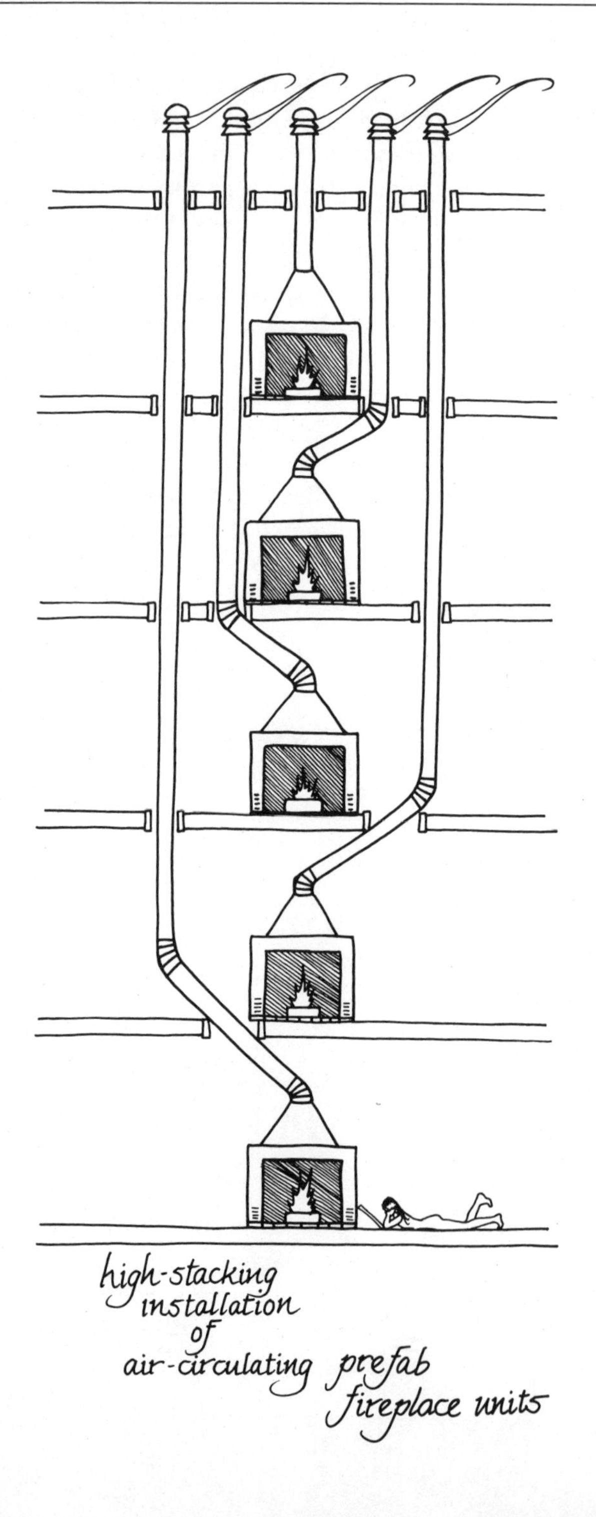

high-stacking installation of air-circulating prefab fireplace units

the ceiling. The other version is similar to the superinsulated standard variety, but it too has vents built in; air circulates around the multiple-layer firebox, comes out on top and is piped up to the ceiling.

Well, here again we have a blast of hot, dry air delivered at the ceiling where least needed. Worse, the fireboxes are standard wide-damper, deep firebox fuel-wasters. But worst of all, we discovered that they can't hold an efficiently banked fire (and any brick-and-mortar fireplace can do that). But all that air circulating around the thin steel sucks the firebox of its warmth —particularly if a system of blowers is installed, as I've heard proposed. True, the circulated air warms your ceiling, and perhaps the upstairs bedroom floor before it circulates around to be sucked out the flue, but only if you keep a great, roaring fire. Let it go to banked coals and shut down the damper and you'll soon have a dead fire. There will be a few sparks deep in the ash bank come morning, but the room will be mighty cold.

You see, the coals in an ash bank feed on themselves; they need to remain in intense heat to generate efficient heat, and that means slow heat produced with a minimum of oxygen. Just enough fresh air to keep their cheeks red. The ash bank acts as an air barrier and in a ceramic fireplace the brick is a heat sink, absorbing the warmth and radiating it back out—most of it back at the coals to keep them hot and the rest at you and me to keep us warm. A sheet metal firebox, particularly one with air circulating around to steal its warmth, can't do either as well. A prefab metal firebox of standard design is to a brick Rumford what a tinny little sheet-metal stove is to a big Austrian Styria stove of cast iron and firebrick. No comparison.

Scandinavian Prefabs

Just recently, some of the people who import Scandinavian stoves have come out with something that is a cross between a stove and an efficient fireplace. It is, I suspect, a prefabricated version of the traditional Swedish hearth, which just naturally opens out at a 45 degree or larger angle, the best for radiating the maximum amount of heat into most rooms.

These are big, wide fireboxes of efficient stove design (with smoke baffles and sophisticated damper arrangements much like the stoves we've discussed). They are made of cast iron and are designed to be set within a heating chamber, a sort of second skin. Air within the chamber is heated by the fire radiating out through the cast iron and flows, or is blown, back into the room. Or, since these systems approach a good stove in heating efficiency, into more than one room through the appropriate conduits.

The sellers claim these units heat at 20 percent efficiency, compared to 10 percent for a regular fireplace and up to 60 percent for a stove. Well, these designs weren't out when Louise and I put in our wood heater and I'm not so sure we'd have installed one if they had. Fireplace to me is spelled Rumford, and the Combi stove we do have puts out enough of a cheery fire in the

Jøtul System 15 stove-fireplace

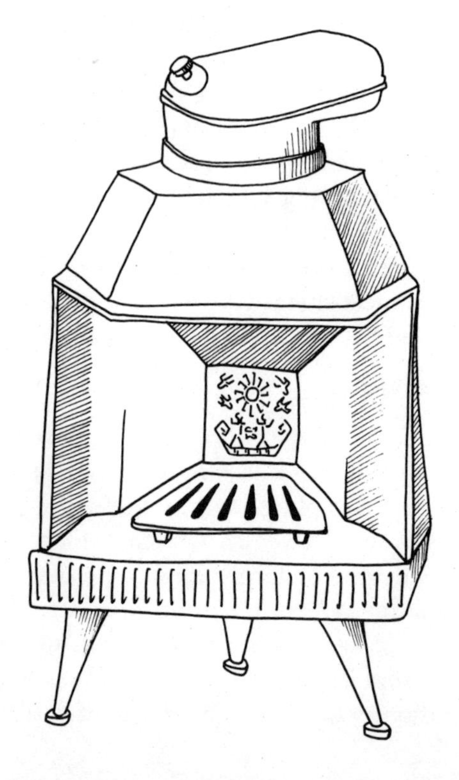

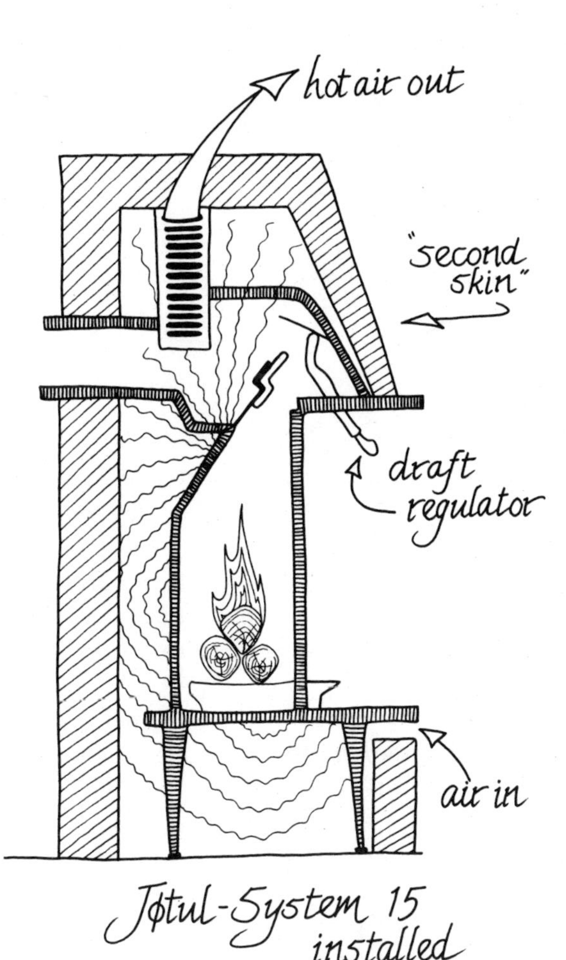

Jøtul-System 15 installed

evening, but, more important, it seals up at night into a truly efficient airtight, spark-free stove. Still, if our dream of cutting off a segment of Maine woods to get the sticks to build our own log cabin within sight and sound of the ocean ever come true, there'll be a homemade four-flue center chimney poking above the roof peak, even if I have to apprentice myself to Dowd or Howard or one of those guys to learn how to set a level brick. One flue for a wood/oil furnace (oil for when we're away) to provide hot water for washing and for heating the greenhouse, keep the chickens' water thawed and so on. The wood-and-propane kitchen range will go on a second flue, Louise's pottery kiln on a third. But #4 will host a big Rumford to dominate the inner wall and radiate good cheer into the living room. Hope you have one too.

CHAPTER FIVE

Cooking With Wood Heat

Cooking with wood is cooking with skill borne only of experience. I guarantee that the first time you try baking bread with wood heat the loaves will come out as our first efforts did—risen at one end, shrunk at the other, burnt all around on the outside and half raw in the middle. Each fireplace, stove and kitchen range is an individual, its performance a function of the flue's size and draft, the type and quality of wood, time of day and season of the year along with the age and condition of the firebox linings, the length of time since ashes were raked out, and the effect of which design you picked from the infinite variety of draft, grate, damper and ornamental designs that have cropped up over the last 1,000 or so years of cooking appliance development.

In addition, cooking with wood is work; you can't set the timer or some gadget that turns off the heat when your roast reaches the proper temperature and go off shopping. The difference between wood and more modern gas or electric appliances is technical. But the difference between the cooks who use the different devices is in the mind and spirit. You can let the modern stove do the job by regulating the length and intensity of heat output on the burner or in the cooking box. With wood—particularly if the fire is keeping you warm in addition to cooking your dinner—the heat source and intensity is fixed. You've a fire and that's that. To cook properly *you* must regulate *the food,* moving it to hot or cooler places near the grate fire, on the stove top or whatever. Since the fire will vary in output as new wood is added and air supply is changed, the moving job is a continuous one. Even in a good wood range where you can regulate

the flow of hot air and smoke around the oven, you must constantly make adjustments.

Then there's the time that goes into feeding and poking the fire, hauling the ashes, cleaning the firebox and air channels and desooting the stovepipe. Wood cooking isn't for everyone at all times, and as the photos show, our own range has a set of gas burners, a gas oven, and a "salamander" (a broiler) on top. We use them plenty. Plus we have a little superinsulated electric oven made by Farberware that doesn't waste heat—keeps it all in just as the wood oven does. The bread gets baked in the electric appliance on a warm August morning you may be sure.

However, anytime the kitchen needs warming, the wood stove is fired up and we cook most everything on it. There is something about a wood-fire cooker and the time and care that go into caring for it that also demands that you put more time and care into preparing the meal. Meals cooked over a wood fire just naturally have to be worthy of the time and effort that have gone into producing and maintaining the cooking heat. You don't build a fire to warm up a TV dinner. And, wood heat seems almost to demand natural foods—old-style *real* food without the chemicals that go into most stuff in the supermarket.

The Open Flame

Cooking in a modern decorative fireplace, whether of steel or brick, is possible but not very practical. They simply don't retain and radiate the needed mild heat in a workable fashion. Fireboxes are too small, masonry too skimpy. To get a fire to generate enough heat to roast a haunch of meat, you must get it too hot—with a great roaring blaze that wastes fuel and gets so hot you can't stand to be near it. The decorative fireplace can't maintain a decent bed of coals to keep a stew simmering properly either.

I'll assume that if you do cook over open flame, you have a properly large (Rumford or similar design) firebox or an open-front stove (Franklin of 34-inch or larger opening or one similar). Either will keep a good bank of ashes and a low fire to radiate the gentle heat you want for cooking. Admittedly, the

Cooking in the old way starts with the proper fireplace. Gay Pietz lives in a circa 1765 farmhouse in central Massachusetts that still has the original kitchen hearth. Here she stirs the beans in a pot on a two-trammel rig under the cooking crane. Since it's spring, the andirons are positioned so the horizontal log rests are close together to support a "short fire," which produces a low heat that's fine for cooking but not for heating. In winter, the andirons are switched so both the ash bank and the logs extend almost the entire width of the firebox, producing a fire that comfortably heats the center of the Pietz house, even on the coldest nights.

fireplace with a shallow firebox is a better open-flame cooker because it has so large an opening to provide a greater variety of heats and more cooking area of easier access to the cook than any wood stove I've ever seen. But we've done up many a stew in and on the open-front stoves we've had and watched a good many pounds of meat roast to perfection in front of them.

Whatever open fire you have, the thing is to get the firebox well heated, a good bed of coals banked, and the wood gently burning down. If you are keeping a permanent fire as described earlier, you've no problem. If starting cold, light up early in the morning for the dinner meal—preferably the morning of the day before. At first you may want to test the coal bed and heats in various areas of the fire. We test fires out with a little oven thermometer that you can buy in any hardware store for a dollar or so. Usually the old-fashioned roasting heat of about 350 degrees F. will be just a few inches out in front of the opening. A good broiling temperature of 500 degrees F. and more exists right over the coals.

Using the Proper Utensils

Cooking over an open fire is the oldest form of the art and you'll surely want to get in a supply of old-time cooking utensils. No need to spend a fortune for real antiques. With today's growing interest in wood heat, more and more firms are making such gadgets as trammels and trivets, cranes and danglespits. Even the local hardware store sells pot hooks and bean pots nowadays. We've adapted campfire cooking equipment to inside use and many items can be made from readily available modern materials. A blacksmith (if you can find one—more and more are going back to work nowadays) would be delighted to work you up a big cooking crane. If need be, a welding shop can put one together for you.

The Crane

The swinging crane with a cast-iron pot hanging from it is probably the most recognizable accessory. Most modern Franklin stoves have them available, and these can be adapted to most fireplaces as well. They are triangular affairs that swing out over the fire. At the end of the top leg there is a hook so the pot won't slip off when you swing it out over the fire. In small stoves and

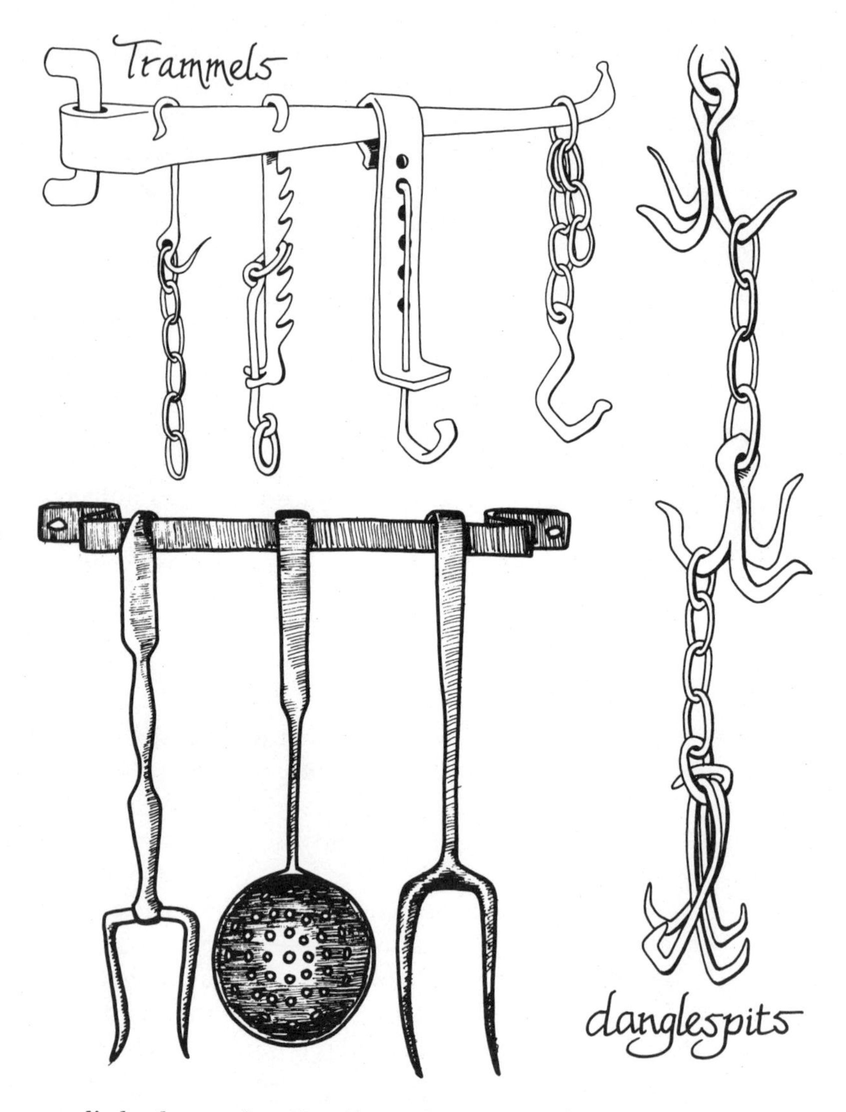

any little decorative fireplace, the crane and a small pot will likely fill one whole side of the firebox and you will have to move coals around and do a lot of pot swinging to keep the heat correct.

In a big walk-in or Rumford-design fireplace, however, you've got the proper amount of space to set up a real colonial cooking hearth. Cranes were usually located at the back and left side of the large walk-ins and were of a length to swing out and have the hook about in the center of the fire bed. They were also placed high up in the firebox. Very large pots could be hung directly on the hook of the big back-wall crane. But for smaller containers, there was a variety of ingenious raising and lowering devices, called trammels. Roasting spits came on link chain; height was adjusted by hanging the spit and roast on whichever link provided the best cooking heat. Pot hooks were adjusted the same way.

A new or servicable old wrought-iron trammel would probably be hard to come by. But you can easily get a length of steel link chain. Just use some stout bolt cutters to snip the middle out of one side of the lower link. It will serve as your pot hook. You can leave the rest of the chain as is and adjust the height by raising and lowering the whole device. Or cut the sides out of all the links and have individual **C** hooks that can be added one at a time.

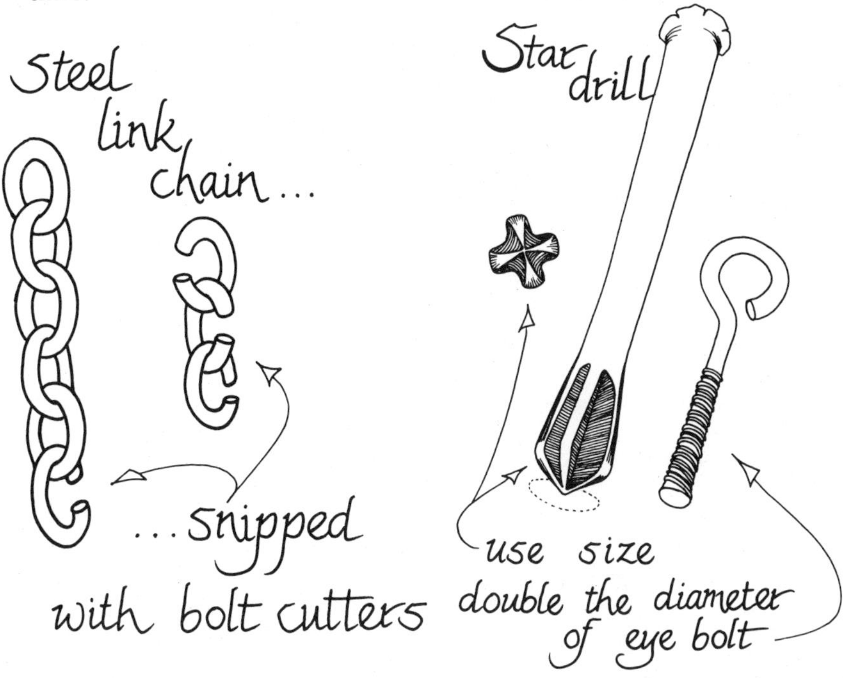

Installing a Crane

Putting a crane into a steel-sided fireplace (if you must) is simple if you are putting the whole thing in new. Just get a crane made for the unit, drill holes and bolt it on before you frame the fireplace in. You can do the same with cast-iron stoves. You'll

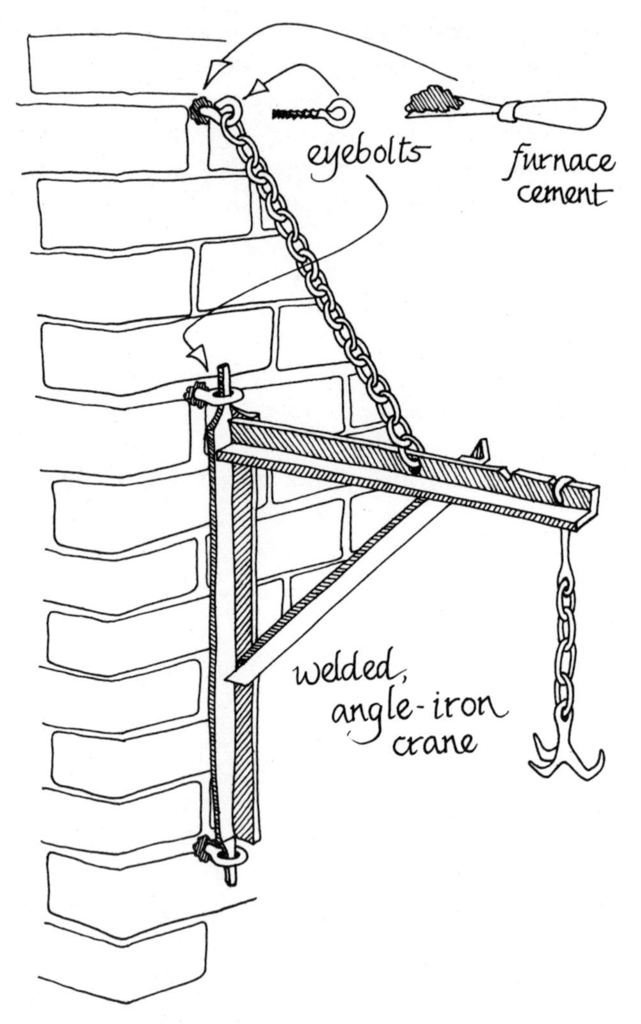

need high-speed carbon steel drill bits and a good power drill for either job.

Installing a crane in a ceramic firebox is more of a job. First, get a "star drill" of the proper size (twice the diameter of the bolts or pins you will use to attach the crane). If using a manufactured crane, get bolts that will go through the holes in the frame, at least an inch-and-a-half long. If using an old crane or a custom-made one, get eyebolts to fit. One possible design of a crane welded from angle iron is illustrated.

Mark your holes—in mortar joints between bricks is best and if making your own design, arrange for bolts to meet with joints rather than bricks. Hammer away at the star drill, turning it a bit after each blow. In time you'll wear a hole in the masonry. If drilling into soapstone, use a power drill with a carborundum masonry bit as the stone is too fragile to hold up to a star drill unsupported. You can use the same rig on bricks and mortar, too, needless to say.

Now turn the bolts in, with stove cement packed into the holes around them. Don't use the lead or plastic expanding plugs for putting bolts into this masonry, as a hot fire will melt them out. In the three-bolt arrangement illustrated, you would want to put the bottom two bolts in, let cement dry, then put on the crane before attaching the top bolt. If you're using the upper supporting chain in our design (really necessary only in weak brickwork or for really big cranes to carry cauldrons), put this in last, as you want the chain to be taut enough to provide support and you won't know where the topmost eye bolt should go till the crane is up.

Cast-Iron Cookware

The only food containers that look right over a wood fire are of cast iron. And, fortunately, many foundries are again making ironware, some of them from the same molds that turned out the pots, skillets and Dutch ovens our great-grandmothers cooked with. For cooking and baking under your crane, get several sizes of iron cooking pots with iron lids. Some are sold with Pyrex glass lids, which let you see the cooking but which never seem to last long around our place. Make sure the pots have stout steel handles (bails). The best selection of ironware I know of is at the Cumberland General Store, Route 3 Box 479, Crossville, TN 38855. For three dollars, they sell a mail-order catalog containing over 250 pages of tools for self-sufficient living. Their instructions on how to properly season ironware are the best I've seen in print, so I'll take the liberty of reprinting them: "Always rub your new cast-iron utensils, inside and out, with animal fat (lard for instance) or vegetable oil. Put in HOT oven and heat. HOT! Do this several times. (You might want to do this while doing other baking, so as not to waste fuel.) When you get your utensil 'cultured and cured' properly, it'll put Teflon in the shade!!"

Seasoning Cookware

In a fireplace, of course, you season by putting the liberally greased ironware on the banked ashes. The purpose of the seasoning is to open up the minute pores in the iron, let hot fat in, then burn it on to form a glaze that acts as a barrier between food and the iron. Never scour an iron pot; if food sticks, it's better to toss it on the coals or bake it in the oven to burn the food to a char. Should the pot rust anywhere or should water-cooked foods have an iron-y taste, repeat the seasoning process. Seasoning is smoky and smelly, but essential.

In time a black coating will build up on both the inside and the outside of well-used iron pots. On the inside the coating will be thin, hardly perceptible in the middle and lower pot. At the inner top and all down the sides, it will be thicker. Nothing but well-charred fats and oils, perfectly sanitary, and as the folks from Crossville say, it'll put Teflon in the shade.

By the way, if you've tried ironware on modern stove tops and ovens and found it unable to take and hold a good seasoning, don't feel bad. So have we. Part of the problem is that with a gas or electric burner or bottom-fired oven, the iron bottom gets too hot too quickly, throwing the seasoning into a water-based dish. Cook slowly. Louise also suspects that acidic foods remove seasoning, so stews containing tomatoes go into ceramic pots.

You should never have problems of losing seasoning over an open wood fire, though. The gentle, overall radiation from a good fireplace will heat the utensil slowly and thoroughly. In a good firebox, the sides of a pot will simmer almost as much as the bottom. What distance to hang a pot from the fire is something you can learn only through experience with your own equipment. A candy or deep-fat thermometer is probably a good investment at first, particularly if you want to use the marvelous capability of ironware to keep an even heat for deep-fat cooking of French fries and the like. In short order, though, you'll be able to do wonders unaided with the crane, your ironware, and the fire.

I'll not offend you by telling you in great detail that you fry bacon in a skillet, make stews in a stew pot, cook chicken in a chicken fryer and fish in a fish fryer, should you want to buy all the appliances. Or that you can do it all in a single flat-bottomed skillet. Get what you want to cook what you want. Pots with bails are about all that work with the crane. Handled skillets and the like need some bottom support, and I prefer to equip a fireplace with two or three log rests. You can scrape coals as wanted out to the front of the firebox, set skillet or pot on the log rests and let it cook away. The log rests are cast iron and effectively transfer additional cooking heat from the fire banked at the back of the firebox. In most stoves, you are restricted to the smallest sizes of pots and skillets; in a big fireplace, you can brew up some really big stews.

Baking, Roasting, Broiling Over an Open Flame

I suppose that a Dutch oven is so called because the Dutch bake in it. Louise isn't Dutch, but she turned out plenty of round loaves back in the big walk-in fireplace we used to have in Pennsylvania. For the novice an oven thermometer is a must. Put a thermometer inside your biggest Dutch oven on the crane and move them up and down, in and out through several day's wood burning. You'll soon find out what combination of pot positions and fire produce and maintain your preferred baking temperature. Louise always used two big six-quart ovens. Loaves were let rise to the proper size in one placed out in front of the flames where it was warm but not cooking hot. Then she'd bake the first while another was rising for the final time. This alter-

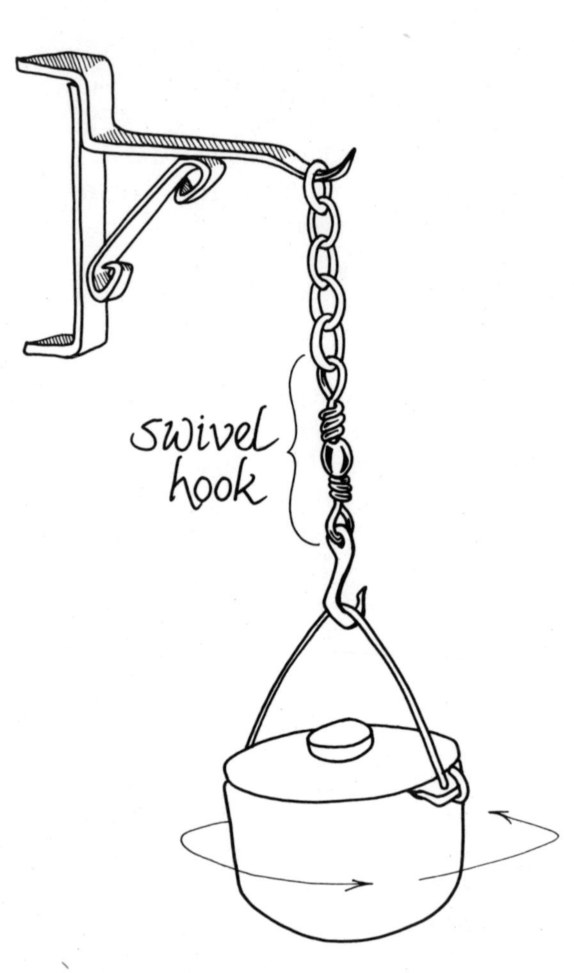

nating of ovens would go on as long as she needed.

With a deep walk-in fireplace, the bricks all around radiate enough heat to keep a Dutch oven quite evenly heated, though the pot and bread usually got turned every fifteen minutes during cooking. With a shallow-boxed Rumford, all the heat radiates into the room—fine for heating, but not for baking. Cast iron retains heat, and you could probably arrange a swivel hook and keep turning it every minute or so. I can imagine a motor or spring-run gadget that would turn it for you. Easier is to erect a reflector—like a giant camp oven. (One of which you can certainly use for quick breads such as biscuits.) But for slow-baking yeast breads, it's best put up a regular reflector—make one of shiny sheet metal as illustrated. Again, trial and error alone can

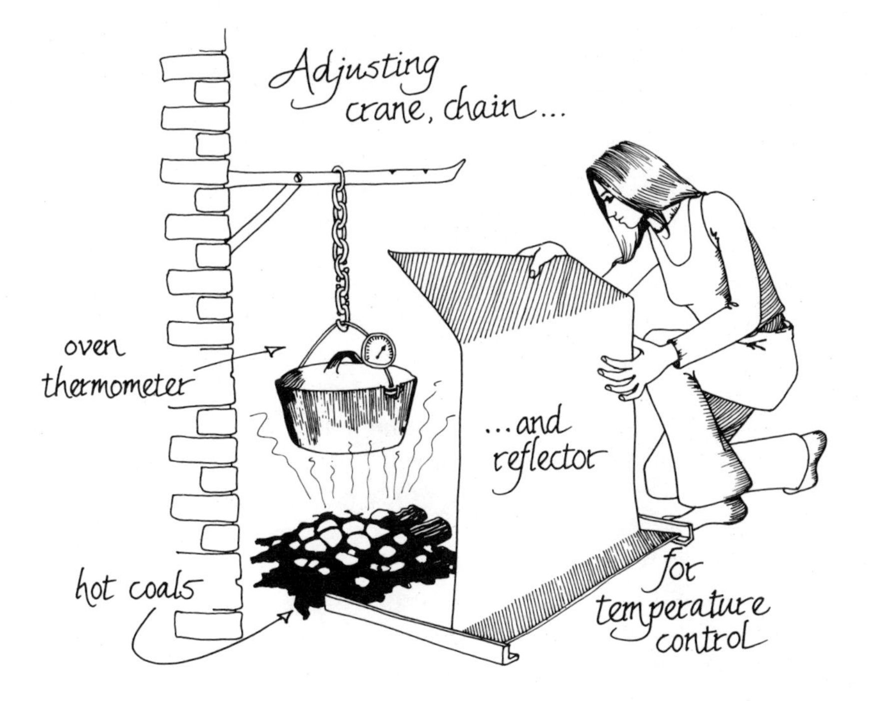

tell you what position is best for the Dutch oven and reflector. Here's Louise's favorite recipe for Dutch oven bread.

Louise's Dutch Oven Bread

Arrange fire, pot and reflector to maintain a good hot 400 degrees F. for an hour per loaf. We found that this heat in our fireplace was in the corner about a foot from the back and left side of the brickwork, the pot a bit over a foot above new, hot coals under ash several inches deep. The coals had to be renewed from the heart of the fire every quarter or half hour, and the reflector placed from one to two feet away. A small oven thermometer was cinched over the bail on the side away from the fire. Checks every few minutes tell if the heat is in the 400 degree F. area. Adjusting the reflector out gives you 20 degrees F. or so of cooling control. A small shovel of red coals pressed under the ash and the reflector moved in gives somewhat more of an increase in heat. You may want to use a camp or hearth pot oven with legs and a lip around the lid top to add coals.

For each large, double-sized round loaf, mix one package of dry yeast in warm water containing a few pinches of light brown sugar. Scald two cups of whole milk (bringing it to not quite boiling) and while it is still hot, add two tablespoons of light brown sugar, two teaspoons of salt and one tablespoon of sweet (unsalted) butter. Mix well, let cool to lukewarm, then add the yeast, stir again and combine with five cups of unbleached white flour and a cup of stone-ground whole wheat flour. Here you can also add up to a half cup of your favorite grain supplement—wheat germ or any number of seeds and nuts. Stir till moderately firmed up, then turn onto a floured surface and work in an additional half to three-quarters cup of white flour as you knead to work the gluten out. Knead until the surface of the dough is smooth and has a satin shine to it, but no more than 10 minutes. Lightly butter the inside of the pot and the top of the Dutch oven. Form the dough into a ball and put in the pot. Cover the pot with a wet towel and let rise out in front of the fire for up to two hours or till it has doubled in size. Now punch the ball down lightly, turn it over in the pot and let it rise and double again, up to another hour. Punch it down lightly once more, turn in the pot and let it get its wind for a quarter-hour. Then put on the top slightly ajar and put the oven on the hook. (If not using a reflector, leave the top off.)

The oven must heat up so the inside has reached the correct temperature, then it must cook with the top off for an hour or more. Only experience will tell you how long this will take in your own kitchen. It varies with size and thickness of the pots, with the shape and heat of the firebox, the reflector and much else. But we found that if the bread is burnt outside and raw inside, the heat was too high and not long enough. If burned on the bottom only the pot was too low or the fire not well banked enough. If burned on the top and not the bottom, the reflector was too close. That was in our fireplace. You may have better success by dividing the recipe in two and cooking two smaller loaves in smaller ovens. Perhaps you will find it best to preheat the pot and after the quarter-hour rest, pop the dough into the already cooking-hot oven. Be sure to leave the top on but ajar, until the bread is fully rerisen. That way the bread can heat evenly for the in-oven rising period but gasses released by the yeast can escape. You'll simply have to experiment, and be prepared for some initial disasters if your luck is as bad as ours was.

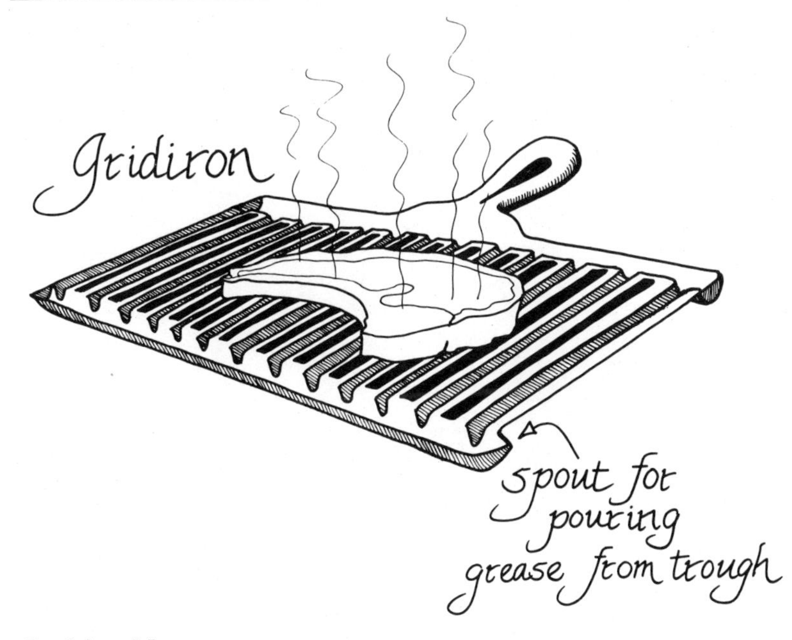

Cooking Meat

Baking bread in a Dutch oven takes a Louise-like patience—and willingness to suffer through a lot of burnt crusts. But the foolproof glory of open hearth cooking, that even yours truly can't ruin, is grilled or spit-roasted meat. If you char the outside of a steak it just adds to the flavor. Or if your slowly turning chicken or pork roast looks as though it will brown before it is cooked through, you just move the roasting spit farther from the fire.

The term to *grill* meat comes from the utensil used in the old days to cook thin cuts of tender meat: a sheet of metal with slits cut in it in the shape of a grill grate. *Gridiron* is another name for it, a word now used mainly to describe a football field, with its alternating stripes and open patches of grass. A proper gridiron is made of cast iron and the only ones being made these days that I know of come on hibachi pots—the little Japanese-style charcoal cookers. Wire grates or such used on home barbecues would work for a time, but would warp if used the way the gridiron should be. We bought an extra foot-square hibachi pot grill that fits just right between the log rests in the stove.

To grill, I simply rake hot coals to the front of the fire, put the grill on the log support till it has heated to smoking hot.

Then the steak goes on the grill and is left over the hot coals till the part of the meat surface touching metal is burned black and the open part is well-seared. Then, before fat can begin falling onto the coals and catch fire, I cover the fire with warm ashes. The steak is cooked five to ten minutes to a side, and I'll tell you, no backyard barbeque over charcoal briquettes can hold a candle to the flavor. A few splinters of hickory wood thrown on just before the coals are covered will add a genuine hickory-smoke flavor. (Be sure not to use any evergreen or birch wood for any part of the broiling fire, as the aromatics can add an unpleasant cast to the taste.)

You can finish the dinner by baking potatoes—whether wrapped in foil or not—for about an hour in the cooler part of the ash bank and cooking up your vegetable in the bean pot on the crane.

The Roasting Spit

Now, the most spectacular aspect of cooking with wood heat—the roasting spit. You can still find rotating spits before the huge fireplaces in English manor houses and their American counterparts. The supports are huge affairs, capable of holding a dozen or more horizontal spits before the flames, and the spits are turned with complicated chain-and-gear or rope-and-pulley devices operated by servants. These spits could hold entire steers for a day or two of cooking, as well as enough fowl to feed dozens of people at once. Few of us live on that scale anymore, and modern roasting spits are geared to smallish roasts and a chicken or two. My favorite modern spit, though, is a true Roasting Jack designed by David Howard to complement the Rumford fireplaces he builds in his post-and-beam houses. A big one could hold a half-grown hog. It operates by gravity; you crank a more-than-200-pound weight up to the roof peak, and as it descends slowly, it turns the spit and the dinner on it through a pulley-and-clockwork arrangement.

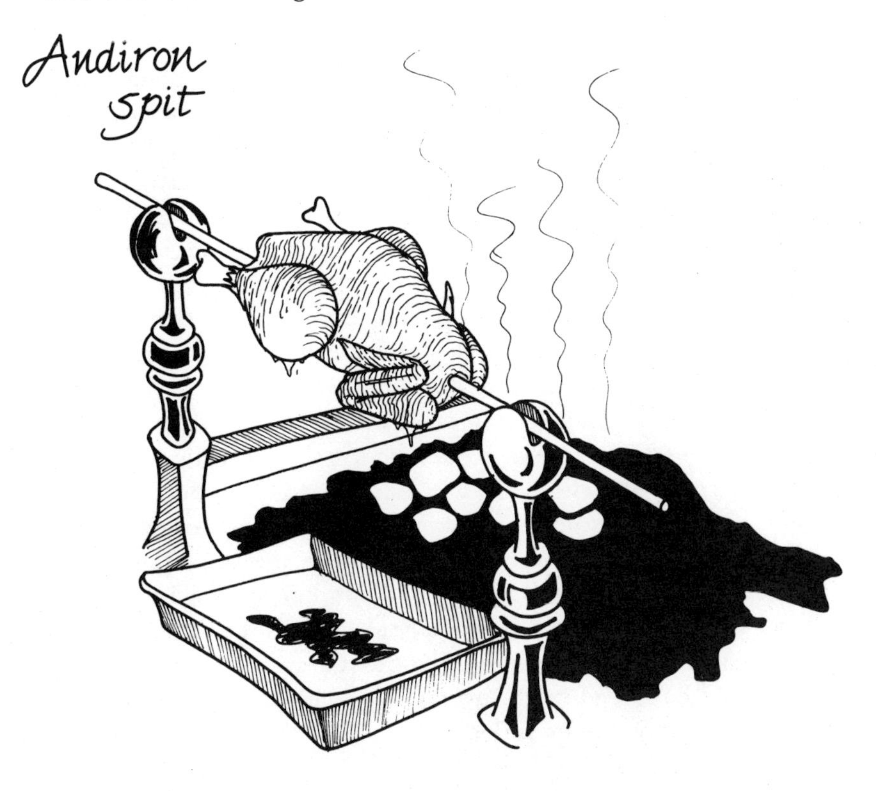

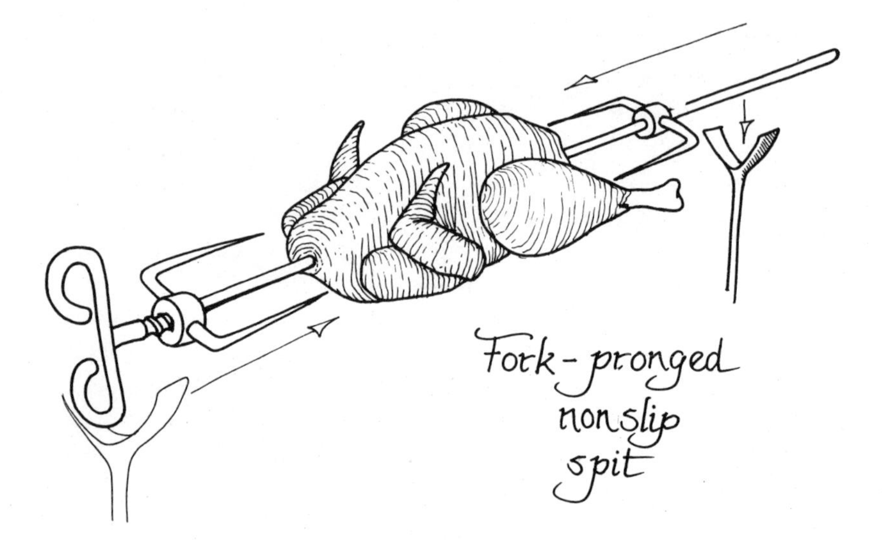

Our own spits have been a bit less spectacular. I arranged a simple spit by banging two notches in the hollow metal spheres on the tops of a pair of old andirons. The bird or haunch of meat was skewered on a length of steel rod, which was suspended between the andirons. The reflector was pushed up and a drip pan put under the meat. It wasn't a revolving spit; the meat had to be turned frequently and the drippings ladled up to baste. But it worked.

As I'm sure you've found if you ever tried to operate a rotisserie of any sort, it is nearly impossible to get the spit through so the meat is equally balanced all around, like a wheel is on an axle. One side or another will always be heavier and tend to flop, as it comes over the top, often working the whole piece off the spit. You can hold it in place with cooking fork props. But the best remedy that we've come across is the meat holder found on modern rotisseries, a pair of two-pronged forks that ride along the spit and have thumbscrews that can be tightened so they'll stay fixed and keep the meat centered on the spit. There are models to go into kitchen ranges, and outdoor cooking units, and they can be adapted to your open hearth. Ours is a little electric-powered one from a neighbor's defunct outdoor cooker. I rigged up the old hood and motor over a dripping pan to sit in front of the fire. The hood helps reflect heat and makes a dandy cooker.

Stove Top Cooking

The essence of stove top cookery can be pretty well summed up in one sentence: build a hot fire and move pots around on the stove top till you find the temperature you want. The steel or cast-iron top will be hottest just over the heart of the fire, in the center of most Franklin and other box-heating stoves, under the oven or firebox in a kitchen range. Most cooking stoves and many heaters have holes in the top fitted with lids that can be removed to expose a pot directly to the fire. Ranges usually have at least one three-part lid with a center circle and two surrounding rings so you can adjust for small pots. Our range, built in the waning days of wood heating, has one three-parter and one solid lid on the left (over the fire) and three smaller ones to the right. When there is a good baking fire going, there will be live flame under all lids and the entire cook-top will be hot enough to boil water with the lids in place. Cooking over open lids leaves soot on your pots unless the fire is down to a bed of slow coals. So we remove lids only when there's something to be brought to a fast boil over a low fire or if we are using a round-bottomed pot or a skillet that is warped or dented and can't sit flat on the top.

Using the Proper Utensils

But no cook has enjoyed maximum flexibility in top-cooking till he or she has used a wood stove. The entire stove top is a cooking surface ranging from super-boiling-hot to just-warm temperatures. If using steel pots or skillets, you can literally see the level of boil change as you move a pot around the top. (Cast-iron cookware keeps its heat too long to be very good for dishes requiring instant changes in temperature.) The top is the only

three-part lid...
...for small pots
...and smaller

true self-cleaning cooking surface I know of. Any spills burn to carbon in short order and are just swept into the fire. If you like, you can fry directly on the iron top, though we prefer to use iron skillets. And the best hotcakes known to hungry man come from the big soapstone griddle that's let warm up on the heating fire overnight, then heated to hotcake temperature—so a drop of water just skitters—over the breakfast fire. You never use fat on soapstone; just mix up the favorite skillet bread recipe and bake right on the stone. It never sticks, never has to be washed and will last for eternity. (The stone will split in time, however, a natural reaction to repeated expansion and contraction. The metal ring around the stone is to hold it together when the expansion joint does appear, as well as to provide an anchor for the handles.) Several mail-order houses sell the griddles, or you can order one directly from the maker, The Vermont Soapstone Company, Perkinsville, VT 05151. The price is hefty—in the high twenties a few years back, but if the stone is given proper care, you and yours won't have to grease another griddle for the next several generations. You can grill meat on the stone, but then it will have to be washed, and grease is bound to run down under the copper ring and end up dripping on the floor. Once it splits, let it cool before you remove it from the stove. Hold by both handles. It is best to store it flat, not hanging. Don't drop it!

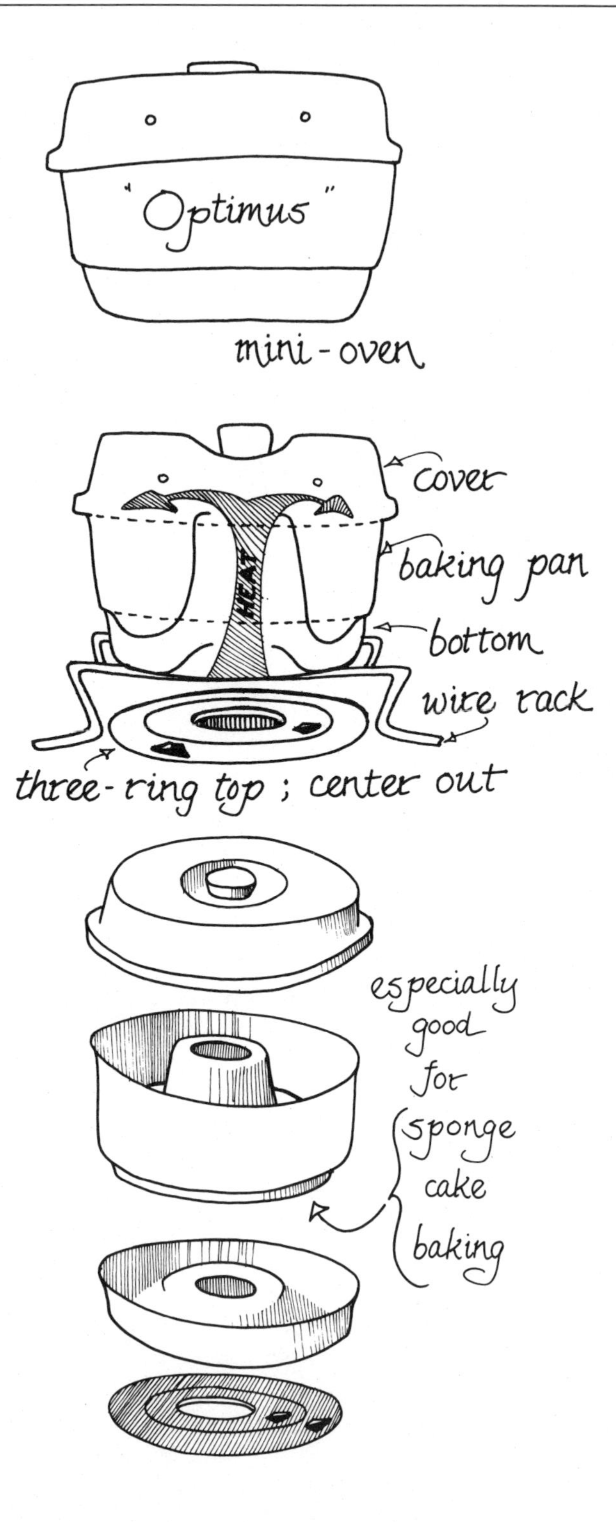
"Optimus"
mini-oven
cover
baking pan
HEAT
bottom
wire rack
three-ring top ; center out
especially
good
for
sponge
cake
baking

Another stove-top bread baker we like for quick breads is a version of the oriental volcano cooker. Ours is from Sweden, made by Optimus and sold by several camping equipment manufacturers. The diagram shows the principle. The cooker was designed for an open flame, but by removing a small cooking lid on the wood stove, you get an even better all-over heating effect. Breads turn out brown and crisp top and bottom.

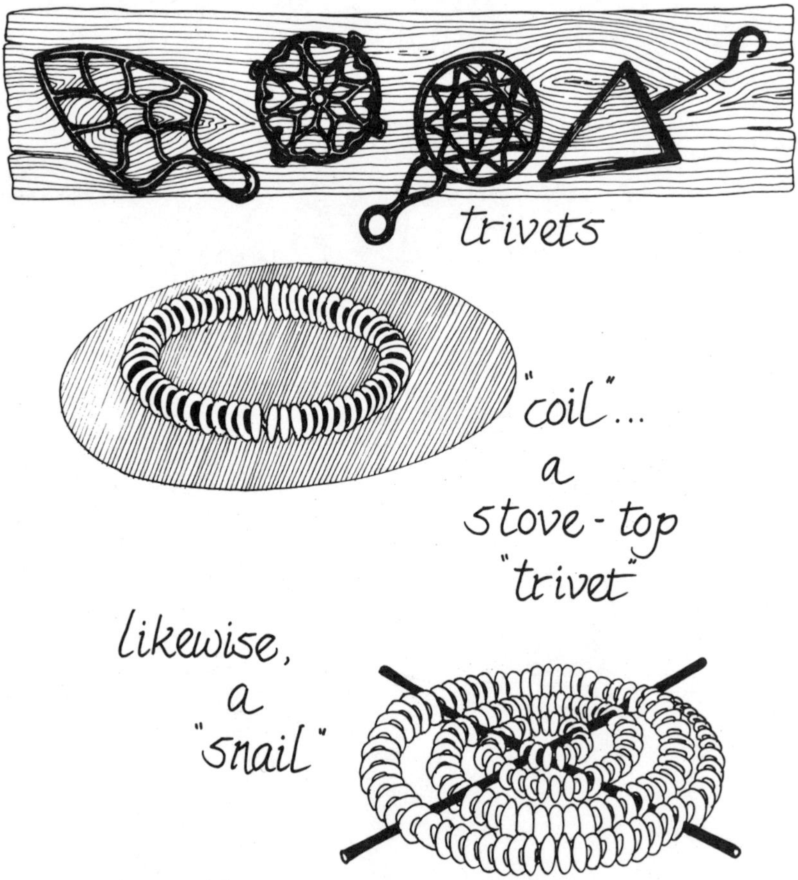

A gadget that Louise finds useful is a trivet, a cast-iron plate, usually of fancy scrollwork design, set up on three legs. It helps prevent hot pots from burning the dining room table. But it is also useful, placed on the cook-top, for keeping dishes hotter than they would be up in the warming ovens, but not cooking hot. Another helper in regulating heat on the top is a set of coils. Just as the name implies, they are lengths of steel coil, like an old-time screen-door spring, twisted together at the ends and pulled out into as near a circle as you can get. Over a coil a pot will simmer slowly without the dish sticking or burning, as it might right on the hot cast iron. You'll get more heat to the pot with a snail, a longer spring turned into a flat spiral and held flat by two lengths of stiff wire running through the coil in an **X**-shape. None of these devices are sold anymore, but they are easy to make from door springs and clothes hanger wire.

The Wary Cook

A few cautions are in order for the novice stove-top wood cooker. First, there is no flame or glowing element to tell you the stove top is hot, and it will retain enough heat to burn you even if the fire is nearly out. I can tell you that the first time I put a palm on a unexpectedly hot stove top was my last. I hope you have more sense than to try it even that one time. Fortunately, if the top is searing hot—capable of burning you seriously, it will

radiate heat far enough to warn you. But we never let the kids roughhouse anywhere near a hot stove and whenever they insist on helping with the meal preparation, they wear sturdy potholder gloves.

Also, *all* of whatever utensil you're using on the top will heat up, including wood or plastic handles of the sheet metal skillets, for example, that would stay cool over a gas flame. You can singe a hand on one. We've found that Pyrex glass pots aren't best for wood stoves. A modern sheet-metal range is tinny enough that a coffeepot that is dropped will bounce enough to stay in one piece. Not so on iron ranges. Also, we've had the pump stems of glass percolators break inside the pot when it was empty and let stay on the hot surface. It just got too hot, I guess. No matter, because wood-cooked coffee tastes best when the grounds are mixed with a pinch of salt, an eggshell, and water from the big iron coffeepot that sits on the stove back—humming and humidifying the air all winter—then boiled up, let sit long enough to settle, and finally enjoyed hugely, muddy color and all.

Queen of the Kitchen—The Wood Range

A lot of wood heat books I've read rave (and it's perfectly justifiable) about the glories of a wood-burning kitchen range and unmatched quality of breads and cakes baked in their wonderful ovens. (Hot air is not continually run through wood-burning ranges, as it is through gas or electric appliances. The wood oven is closed tight; hot flame and smoke are channeled around it—top, bottom and all sides but the front in the best designs—so the heat radiates from all directions. There is no way to bake a better loaf of bread. Indeed, once you've tasted your first success with the range, you'll wonder how the home appliance industry managed to foist their gas burners off on our mothers.)

Not only were the authorities of little help in teaching Louise and me to operate a wood range, even the 1925 booklet *The Secret of Better Baking* by Mary D. Chambers, (reprinted by and available for small change from the Portland Stove Foundry Co., Portland, ME) fails to divulge all the secrets. It assumes the use of coal, and it just naturally expects the reader to know a lot of information that has largely been forgotten in the past two generations. For example, the booklet assumes that we all know what a spider is, as well as the difference between a dock ash grate, a plain grate and a triangular grate. I still don't understand grates to be honest. The dictionary says a spider is a long-handled skillet, originally with legs to go over coals. And here is what Louise and I have picked up from Ms. Chambers and other genuine experts, along with a lot we learned the hard way.

Shopping for a Range

If you've found the $500-plus and waited your several years for a new cast-iron Princess, I envy you. But, many folks will be

There is no place on earth that's better for rising bread than a wood stove. The place you put the rising pans varies with what is going on in the oven or on the cook-top. Here, it's near time to begin baking, so the oven is being heated up. So, the loaves go down to the cool end, so as not to rise too fast and get all cheese-holed.

old range
modern range

buying used, and the problems of burnt-out grates, missing doors, frozen bolts and rusted liners that complicate the purchase of an old heating stove are compounded when you go looking for a kitchen range. Since hundreds of manufacturers turned out thousands of models, I doubt that there is anyone living who could list everything you should look for in judging an old stove. I know I can't. But if we refer to the cutaway drawing to discuss how the range works, you'll have a fair idea of what to look for.

The Boot

Very first thing, go around in the back. If the stove does not vent directly through the back of the cook top, there must be a cast-iron elbow-like projection, the boot, attached where smoke is emitted and forming a 90 degree upward turn, ending in an oval form to fit stovepipe. This is the part most often missing in the old stoves I've seen. You must have one. Even if you managed to adapt a piece of stovepipe elbow to serve as a makeshift boot, the hot gasses roaring straight out against the turn in the pipe

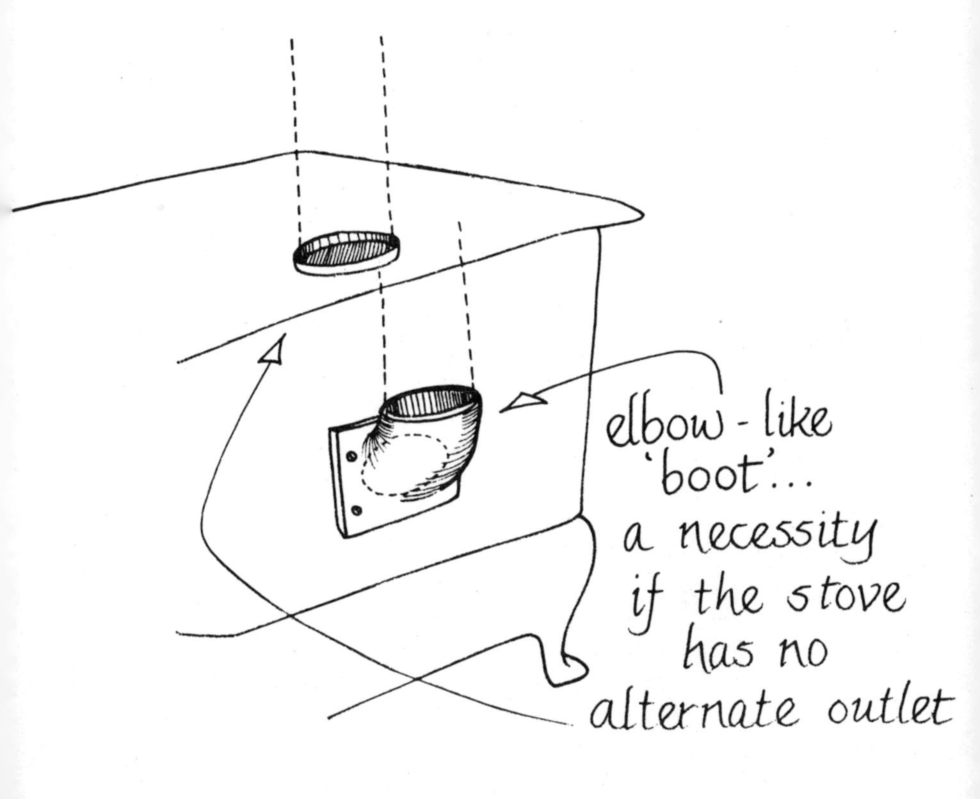

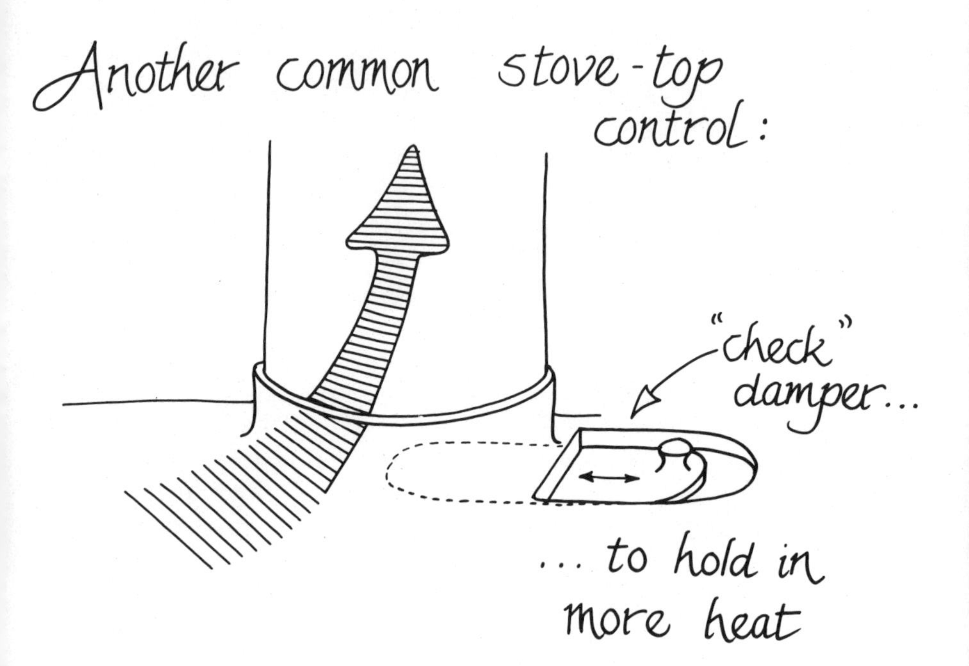
Another common stove-top control:
"check" damper...
... to hold in more heat

installing damper...
if none in boot

when you are baking biscuits would burn it through in no time. You could conceivably have a new boot cast, but that would require someone to make wooden or plaster patterns, two of them to be welded or bolted together, and you'd be better off waiting in line for a new range. The boot may have a damper built in, most likely attached to an operating lever that comes out at the back of the range. This is a convenient thing to have, but if it is all that is missing on an otherwise good range, you can install a regular stove damper in the pipe, perhaps running the operating rod of the damper out through a hole drilled in the sheet metal back that supports the top shelf or warming ovens. Point out the missing part and offer the seller half of what you think he really wants for the range.

The Firebox

Now check out the firebox on the left side of the range. There will be a draft control at the bottom, another at the top usually, and perhaps this hinges out (for broiling). Make sure all hinge pins work, that the draft slides or doors open and close nearly airtight. Don't buy any with damage you aren't sure you

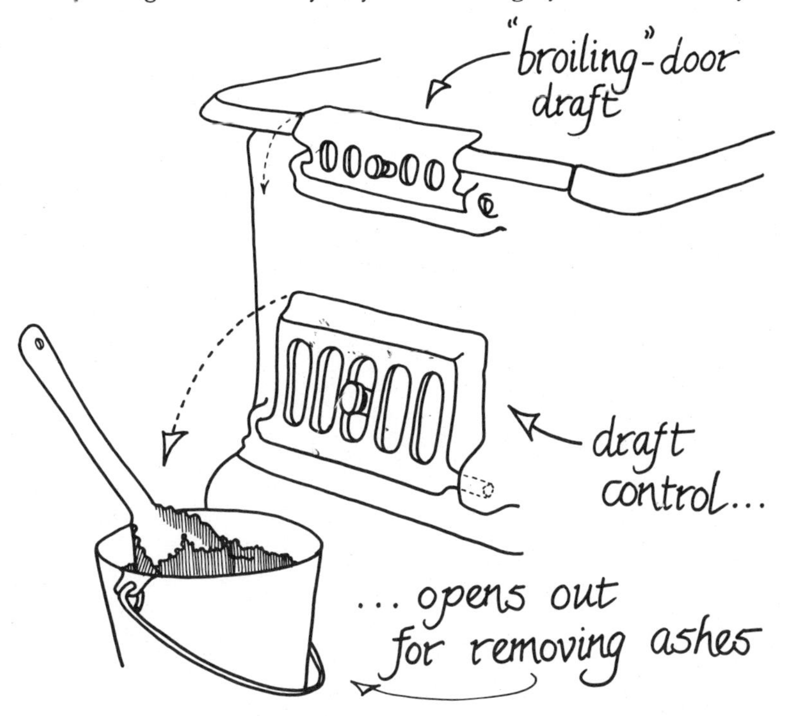

can repair with whatever skills you possess or can hire. The grates halfway down in the firebox side—a grid or a metal rack or a plate over a set of rods that can be shaken or rotated—ideally will be sound and operable. You'll find a firebrick lining all around if the stove was made for coal and cast-iron plates, perhaps removable, if it was meant for wood only. Ours has a double set of liners, one permanent, another of thick iron plates that can be removed. This is a feature of some wood/coal stoves, the extra liner is meant to come out when using wood. We leave ours in to keep the old girl going at least another half-century.

If the grate shakers don't work, it is no tragedy for wood use. You can just push ashes down through the grates with the poker. Revolving or shaking grates are necessary with coal, as the ash and klinkers have to be removed each morning, or else the grates will burn out. It's not the case with the cooler-burning wood. If you find a stove with a firebrick liner though, be sure that the

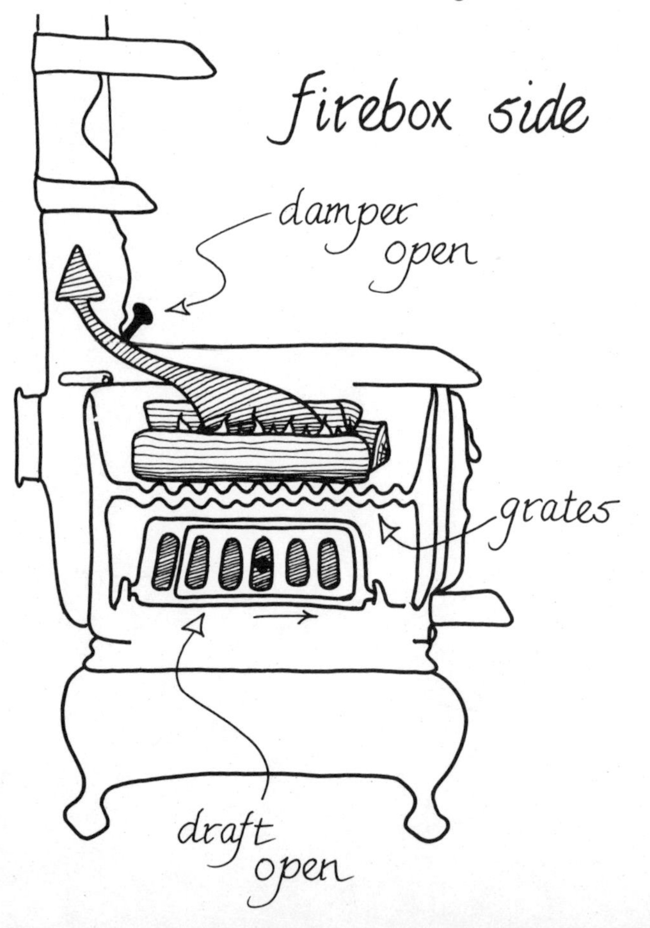

liner is not burned through to the metal, or be prepared to patch it or reline with firebrick or furnace cement. The firebox walls are the most likely part of the range to be cracked or warped. No matter what you're told, no one can weld a cracked firebox so it will hold up for a good time, even if he does use nickel welding rods. And the firebox of a cookstove probably gets the most stress due to changing heats of any part of any solid-fuel heating device. So, check the firebox particularly well. If it can be relined with sheet metal or firebrick, a small crack in the firebox wall doesn't condemn the range to the dump. Or, you can just plan on patching it with stove cement every so often. No big chore. A wood fire gets hot, but not nearly as hot as the coal fires that probably cracked your stove. Warped plates along the firebox sides, front or back should be carefully inspected. Here, as elsewhere on the stove, if you see a lot of pits, the stove has been allowed to rust, most probably it's been out in the weather for years. Triple check hinges, etc.

Is It Repairable?

The older (and costlier) the stove, the more primitive was the casting technology of the time and the more likely you are to have warped plates. For really serious daily use, I just wouldn't buy an antique, warped or not. Any stove manufactured much earlier than 1860 belongs in a museum, I'd say. Too, I feel that most stoves built prior to 1900 just shouldn't be asked to go back to daily work. A lot of folks will disagree, including Richard "Stove Black" Richardson who runs the Good Time Stove Company on Route 112 in Goshen, MA and who knows more about old ranges and most other kinds of old stoves than anyone I know of, myself included by a long shot. Or maybe I should say that I'd not buy an antique range for serious use unless Richard or another expert like him surveyed it and gave it an OK. There are just too many models, styles, and casting techniques that have come up over the years for anyone but a full-time specialist to understand. And, unlike most heating stoves, ranges have several draft controls, integral dampers and baffles that the average person can't understand without a lot of trial and error, much less judge at first look.

However, there will be a lot of folks who just can't resist one of the old black-iron beauties. Well, in an antique don't be put off by rusted or even burnt-out grates. You can replace grates for wood very easily. Coal, no. But a good welder can cut you heat-proof steel rods, fine for wood heat, to fit in back to front or side to side, put in through holes bored in the firebox wall or on legs that go down into the ash pit or any one of a hundred improvisations. Forget the grates, but do worry about the firebox plates. Put back into use after two generations out in the weather, a warped and cracked stove could literally fall to pieces on a first firing. If the removable ash box below the grate is gone, you can have a tinsmith make up another, or you can bend and rivet your own from good sheet metal. Don't let the ashes just collect in the lower chamber; the heat will (usually) burn out the sheet metal bottom in time. Perhaps you can simply line the ash pit with metal and a layer of sand as with a sheet metal heating stove. Check the stove-top and lids for excessive wear or warping. You can probably purchase new lids from one of today's stove foundries as sizes were and are pretty standard. The holes where the lid lifters fit are nearly worn away on a couple of our own 50-year-old lids. I figure that in another 50 years, we or our replacements will have to begin looking for replacements for the lids.

The Oven Controls

Now, remove all lids and as much of the stove top as you can, use a penlight and peer into as much of the ducting as you can. There will be a small door at the back of the firebox, normally operated by a lever located at the range back to one side of the smoke outlet. It has two settings. At *kindle,* the door is open so fumes go directly into the boot; this is the setting used to start fires or for top cookery. *Bake* closes the door, so the heat is pulled over the oven, all around, under and back up into the flue. This control *must* work and all parts *must* be in excellent condition or your oven will never function easily. I suppose that you could rig up a replacement that could be stuffed into the opening through the lids, but I'd be of no help. However, the door and handle parts are usually relatively simple, are bolted on and could be recast.

The air channels over and around the stove will probably bear the residue of the last fires lit in them: a solid packing of fly ash. You should be able to see the top-oven air passage through the stove top. In any good range, the sides and back will be accessible through ports. If you think that you'd never be able to

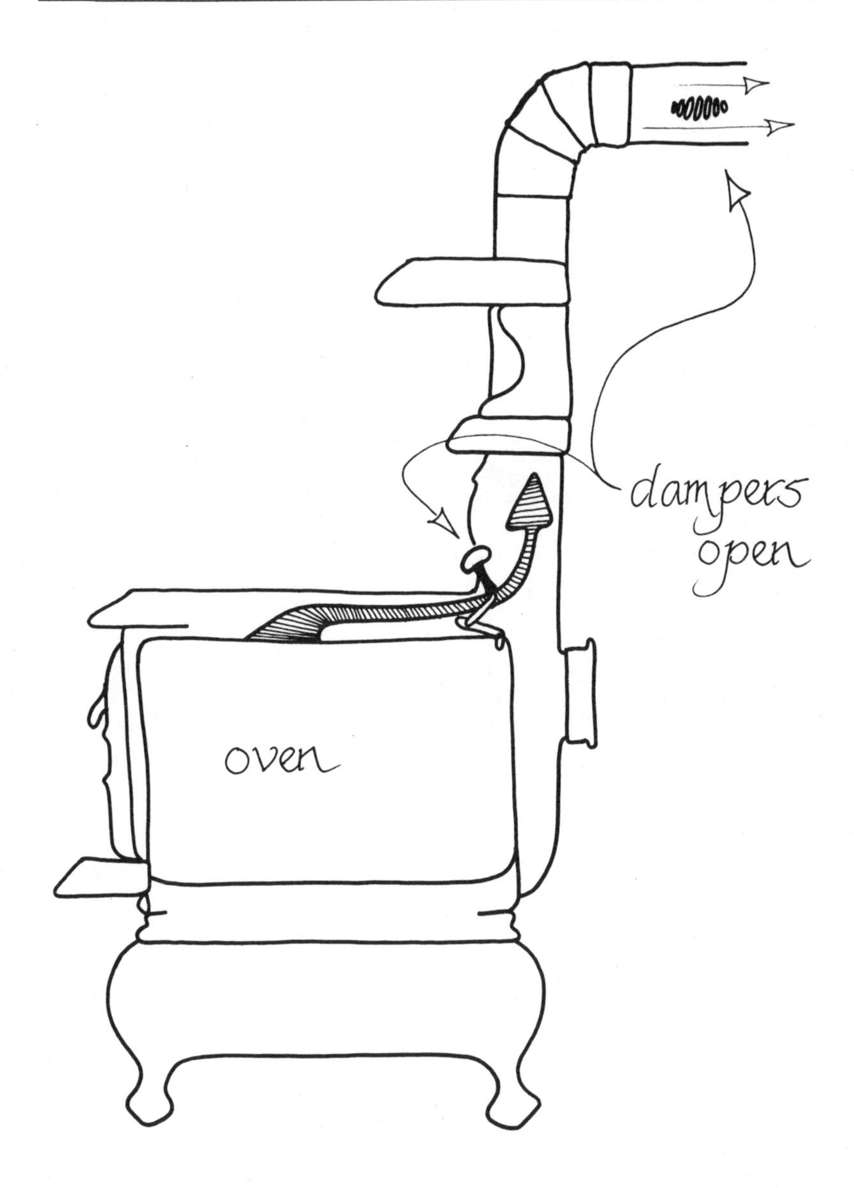

side opposite
firebox

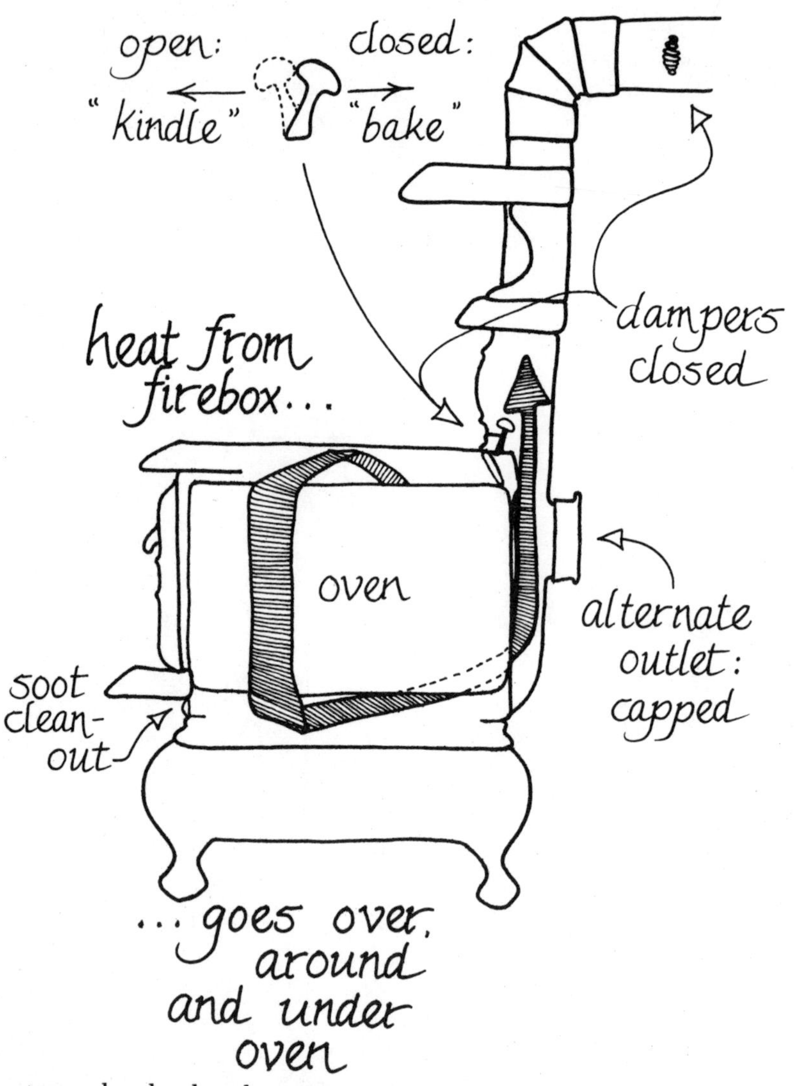

get an ash rake, brush or vacuum nozzle into the crannies around the oven, don't buy the stove.

Finally, check the iron and sheet metal parts for rust. Most of the better stoves had such parts as warming-oven bottoms made for easy replacement with the removal of a few bolts. You'll have to make or have made any replacements, but the job is easy for any shop. If hinges on doors are broken *or* badly rusted, you should look elsewhere. Hinge *pins,* though—if separate steel rods—can usually be replaced if you can get the damaged originals out.

Optional Equipment

Such extras as gas burners, water fonts and the like take extra scrutiny. Since our subject is wood heat, I won't go into gas, except to admit once again that the gas side of our Glenwood is a great blessing come late June. If you are lucky enough to find a range with water heating capability, don't pass up a good cooking range if coils are clogged or the font rusted a bit. They can be easily repaired.

Most water fonts (or fronts) are nothing more than a metal box. If yours leaks, try to remove the whole cast-iron top, rather than just the hinged or removable dipping lid. This is to get at the liner, usually tinplate or zinc. It should lift right out; if it's really in bad shape, you can go at it with a crowbar. (Be careful of the relatively fragile cast-iron outer shell if there is one, though.) If the leak is small, get your Mapp gas torch and braze it shut with the brass rod. Don't use lead solder, which would hold, but would give anyone drinking the water a case of slow lead poisoning. If the liner is shot or missing, write down the dimensions and have a sheet metal shop make you up a new one. Sheet copper is good but expensive. I think I'd just have it made of galvanized steel, (brazed again, not soldered with lead) then have the whole thing given a heavy inside coat of any nontoxic metal: nickel, tin, gold or silver. Silver'd tarnish, of course and gold would cost a bunch (not as much as you'd think) but would last forever. Imagine casually opening up your gold-plated water front to show any gotta-keep-up-with-the-Joneses neighbors.

If your stove has a coil, one or another arrangement of piping to heat water by running it through oven or firebox, it will almost certainly be partly clogged or will get that way with minerals if your water is at all hard. In the old days they used reams to clean the straight sections and let vinegar sit in the curves to eat the clog out there. You can do the same (and can avoid the problem by setting up a rain barrel and heating only naturally distilled, pure water), or you can take the coil to any good heating contractor. Lots of modern hot water furnaces have coils in them to heat the domestic water supply. Ours does, and the coil must be cleaned annually. They open it up and run acid through for a reasonable fee. You can take your coil to the shop and they'll do it even cheaper.

Wood-Fired Ovens

Now, your wood range is home, you've got her cleaned out, polished and spruced up and want to cook a meal. Okay, have a can of stove cement handy. Installation and piping is the same as any stove. No need for a damper in the pipe unless there is none in the range smoke outlet.

We've discussed the damper (the *"check"*) and the oven diverter *("Kindle/bake")*. The way they work and the instructions—the way *"check,"* *"bake"* and all are conveyed on the stove—vary as much as the mechanisms. Some are words molded into the stove top or control handles, others are cryptic symbols or abbreviations. But, you'll figure them out. Basically, you (may) have a damper and (surely) will have some sort of mechanism for directing hot gasses into 1. the flue and, once the fire is going, 2. around the oven.

To make a fire: Open damper or *"check"* control fully, put

diverter on *"kindle,"* open the bottom draft control fully and open fire door or remove stoking lid. Let the flue warm a bit, meanwhile, holding a burning match down in the firebox. If the smoke and flame of the match are sucked out with vigor, we're halfway there. If not, there's an obstruction somewhere between stove and chimney cap that you'll have to find and cure. Lay a regular stove fire of crumpled paper, splinters and kindling. Drop a match in, and once the paper blazes, close the firing opening. Once the kindling catches, add more wood. In a half hour switch from *"kindle"* to *"bake"* and stand back.

If the stove is clean clear through, you'll hear nothing but a slightly reduced draft pull. If the whole machine begins to smoke, you have clogged channels in the oven and a session of excavating is in order, if you haven't cleaned the oven channels already. The clog must come out, and fortunately, these old ladies are not all that mysterious. You may have to drill out a few soft iron stove bolts to open up the clean outs, but the technology

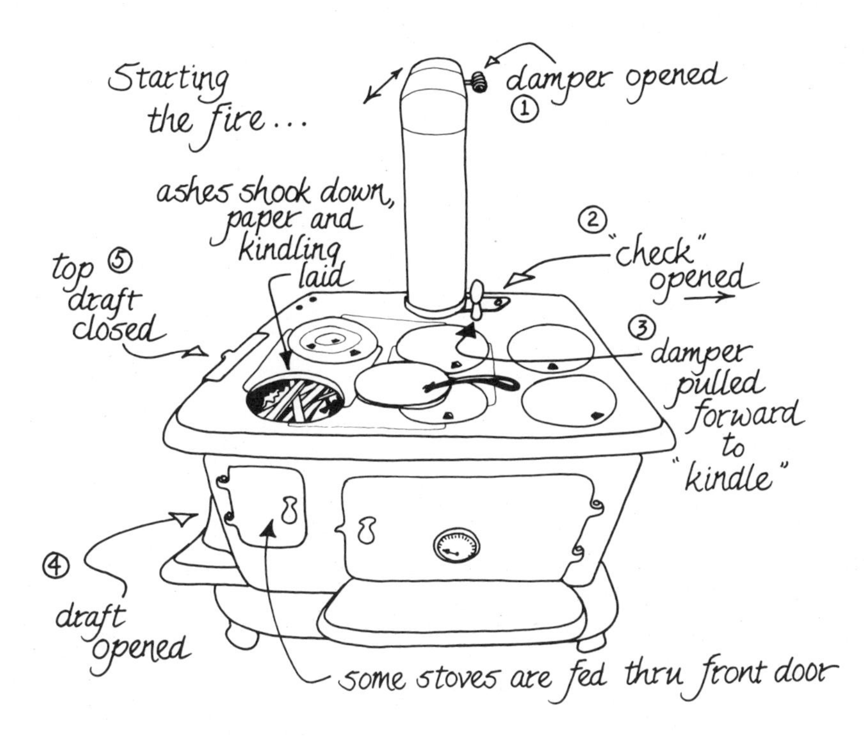

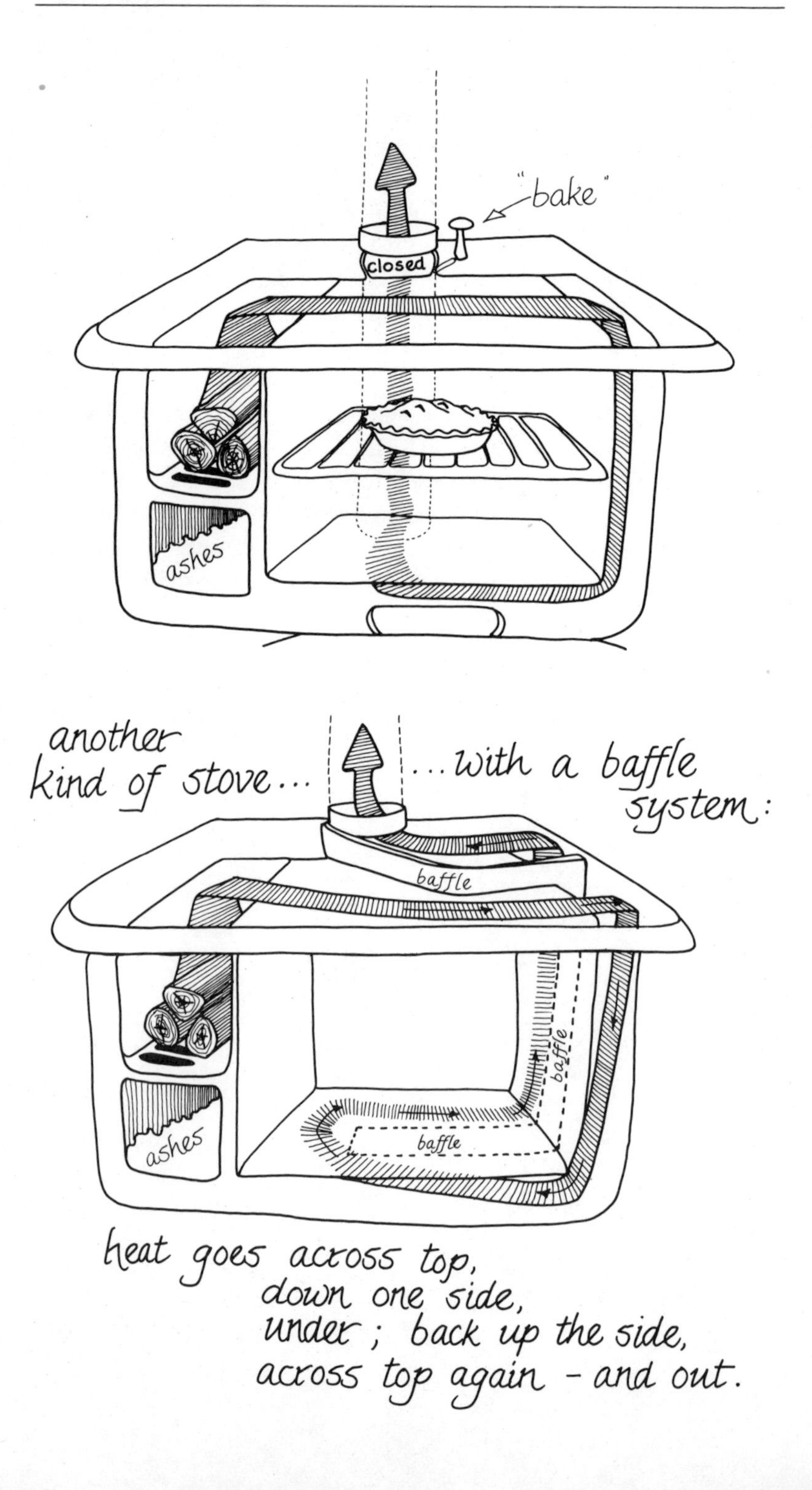
"bake"
closed
ashes
another kind of stove... ...with a baffle system:
baffle
baffle
baffle
ashes
heat goes across top,
down one side,
under; back up the side,
across top again - and out.

Cleaning a wood range is a messy job that invariably gets put off longer than it should. In addition to scrubbing spots common to all stoves, you must remove the accumulated soot from the channels surrounding the oven. The ash rake and shovel are specifically designed to get into all the corners of the smoke channels, but it usually does take some probing around. A quick once-over each month and the fly ash never accumulates. Inevitably though, the newly acquired antique (like this) usually needs more than a quick once-over that first month.

is straightforward and once a good old stove is set up and cleaned, it should be good for the rest of your life and mine, as long as it's kept cleaned out.

I can't begin to tell you how or where the fly ash and soot that's clogging your range may be. Each model has it's own series of smoke channels. You'll have to remove all top plates, bolted or spring-clipped-on ports at top, bottom, ends or back and dig away. The **T**-shaped ash rake we mentioned earlier is an essential tool for getting into the crannies. A stove brush (like a bottle brush, but on a long handle) helps and is good for cleaning out stovepipe too. I found that a vacuum cleaner with a long flexible pipe attached was almost essential to get into some of the harder corners. But, basically, the smoke channels of all wood-fired ranges become clogged with flying ashes and soot in as little as a month's hard use. Once you have yours cleaned of any extensive accumulation, plan a monthly clean out and the hard job will never reoccur.

If smoke spouts from any cracks in stove or pipe, remind yourself to caulk them with furnace cement next time the fire dies.

Gauging the Heat

Most "modern" wood-fired ranges built since 1900 or so have a temperature gauge in the door. It is *utterly* unreliable except as a very general guide to how hot the door is. Moreover, the old-time cooks didn't insist that recipes be cooked at precise temperatures. There were general temperature ranges, and the two gauges on our 1925 range were intended to help guide older cooks into the "new and modern" age of gas, by mixing the old terms with modern temperature numbers. But they serve as well to guide us newer cooks back into the old terminology.

How to attain the oven heat you want and hold it is information each cook simply must learn by trial and error. Louise finds that two small oven thermometers, one placed at the left front, the other at right rear on the oven rack gives the best reading both of general temperature and evenness of the heat. (And wood stoves never heat evenly, thus the accuracy of Jimmy Rogers' 1920s era song line: *"I can smell yo' bread a-burnin'; turn yo' damper down. If you ain't got a damper, good gal, turn yo' bread around.")*

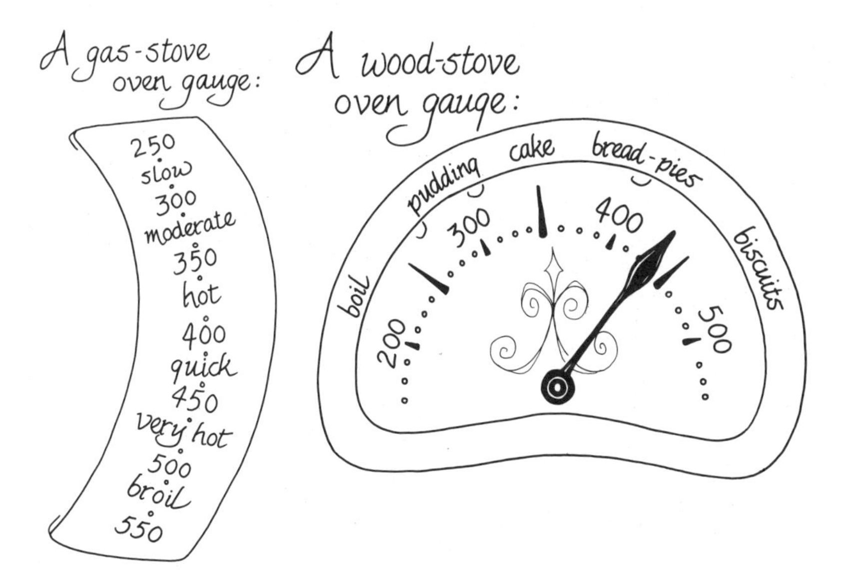

Controlling the Heat

Just as with any wood stove, you use dampers and draft controls, size and variety of wood to vary the intensity of heat. The more air passing through, the more oxygen is presented to the wood, and the faster and hotter it will burn. Generally, the smaller the split, the dryer and lighter the wood, the hotter the fire. *Biscuit wood,* is used to attain the brief, but very hot fire needed for this soda or baking powder bread—as the gauge indicates. Small splits of any wood will work. Small "trash trees" provide the best biscuit wood, if well-dried. A firebox full of poplar or aspen will last just long enough for a batch of biscuits. Two loads of sumac will work in a pinch, and the alder you've weeded out in the spring and dried under cover is good come winter. Cottonwood is good too, but needs a long drying period.

The oven can overheat easily and will with regularity till you get to know it well enough that you must cut down on fuel or bottom draft at the right time—well in advance of the time on the clock when you desire the change to take place. To cool it, restrict oxygen flow using dampers and draft ports as with any stove. Opening the oven for a few seconds works well with some dishes; others such as cakes or bread will fall flat. So, keep a few bricks, flatirons, or a couple of big stones handy. Put quickly into the range, they will absorb heat and cool the oven. How many of

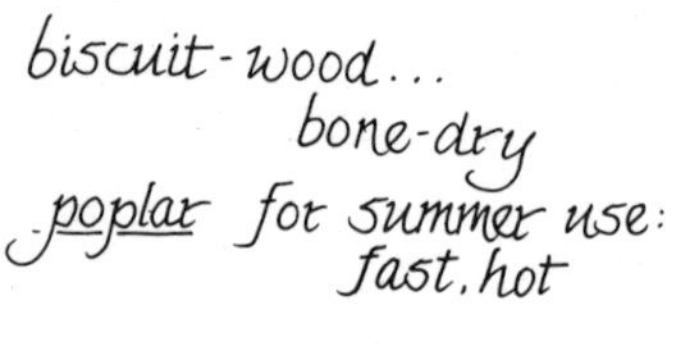

what to use and how long to keep them in is another trick that each cook will learn only from experience.

The brick or flatiron trick is to achieve a quick heat reduction. For a normal two-temperature dish such as popovers or Yorkshire pudding, you can reduce the starting high fire to moderate by opening a large draft over the firebox to let more room air into the stove and wait till the fire dies down to the heat you want for the second moderate cooking temperature.

Here's how Louise cooks the Yorkshire pudding while I'm tending the roast beef on the roasting spit out in front of the open-fire stove.

Louise's Wood-Cooked Yorkshire Pudding

When the roast begins its two hours or so of slow-turning in front of the open stove, the cookstove is put on "*bake*" and filled with a load of good coals-producing wood—hickory or beech or oak chunks.

In the big, copper, egg-beating bowl (the tiny trace of copper that gets into the eggs helps the whites puff up) Louise mixes a scant cup of unbleached flour and a half teaspoon of salt. She then adds a half cup of rich milk or half-and-half and stirs till well mixed. Then in go two eggs, one at a time, the batter being whipped till frothy after each with the big wire whisk. Then a half cup of water is added and the batter is whipped till bubbly, covered with a damp cloth and put aside.

At the end of the first hour of roasting, the pot beneath the meat will contain a good supply of fat and drippings. This is put on the cook top for 15 minutes or so to heat till sizzling. The firebox is filled with biscuit wood and the bottom draft and the damper are opened full. The oven will reach Hot, or about 400 degrees F. by the time the fat has become smoking hot. The batter is poured in and the pot is popped into the middle of the oven. The hot fire is kept going, with more biscuit wood added if needed, for 15 minutes. The pot is given a half-turn every five minutes during this time so it will heat evenly. (Breads of all kinds need a quarter-hourly turn, in our oven at least.)

At the end of the hot bake, the firebox is filled with quarter-splits, pieces of a six- to eight-inch-diameter log that has been quartered. Now the fire is "checked" to reduce the temperature. This can be accomplished in several ways on our stove. The front doors of the firebox shaker and clean out may be opened. So may the top broiling door or the air slots in it. The bottom draft control can be closed, and the damper in the boot may be half closed to the "*check*" setting or closed entirely. Just which measures are used depends on the strength of the draft that day, the wood and all those other imponderables. Usually Louise checks by half-closing the damper, partly opening the upper draft control and closing the lower control halfway. The oven is kept at "*plain cake*" or about 350 degrees F. for another quarter hour or so, baking till the pudding is well risen and golden brown on top. Then we are treated to beef and Yorkshire pudding, with a green salad and hearty red wine, a classic meal harking back to old England's open hearth cookery.

Accessory Uses of the Oven

Broiling with the range is the same as in the open fire, except that you need a long handle on your grill. They still sell dual-handled grills for use over campfires. Build up a strong bed of coals, open the draft and the damper full, and remove the top lids and center divider. The meat can be seared through the broiling door if you have one, or just poked in through the open top. Some old-time cookbooks recommend actually laying the meat directly on the coals, though I can't see how you could avoid getting ashes in the main course.

Once you've had a warming shelf or ovens above a wood-burning range, you'll never imagine how you cooked without them. Plates warm perfectly there, and many cooks just keep the dinner plates in the warming oven when they aren't being used. Bread rises perfectly, salt stays dry and loose, yeast bubbles up quickly, and any dish you want warm but not hot, such as bread or leftover apple pie, warms up in jig time.

Don't neglect to use the towel racks—one or two swing out from the left side of most ranges. If your old-timer is missing

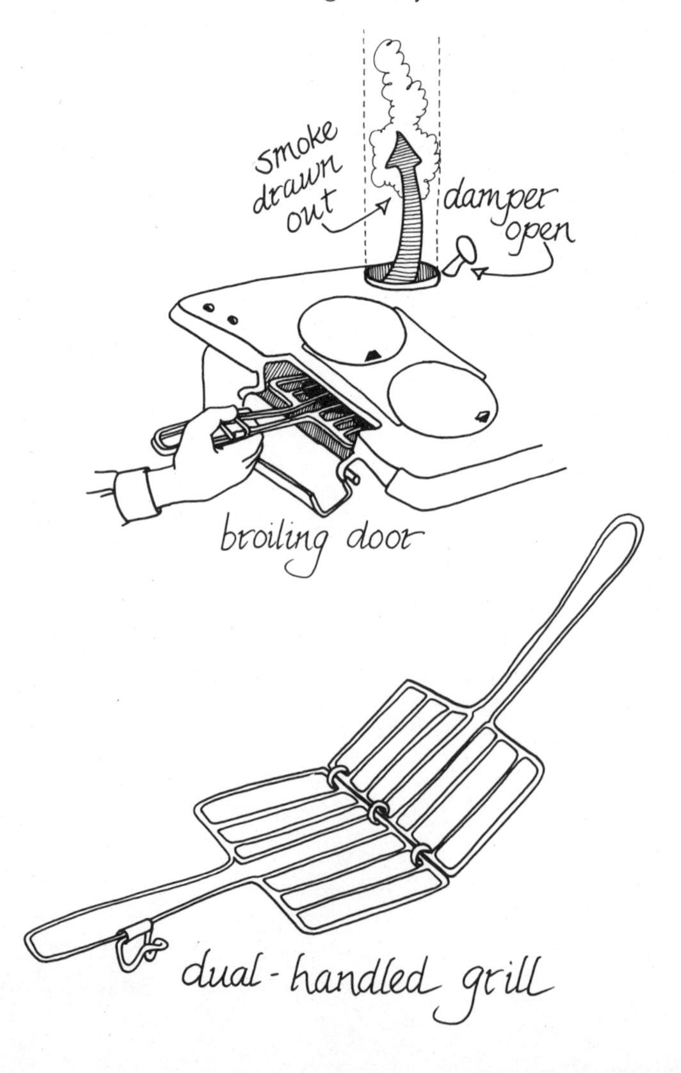

them, you can fashion new ones to fit in the holes in the left-hand rear of the stove top. They are meant to hold the dish towels for drying, but they are fine for snow suits, mittens, and the weekly laundry, bit by bit.

Once again let us apologize for being so vague about the specifics of using the oven. But no two bakes are completely identical. Louise has repeated a huge success detail for detail on a following day and had disappointing results. So if she tried to give you advice on whether to bake your bread on the bottom of the oven, or on the rack put on first, second or top setting, it would be downright pointless. You simply have to try each in its time. But once you master the range, you can be proud of a genuine skill. And, proud in the knowledge that *you* baked your bread, cake or pie. It was all your doing to wrestle that cantankerous machine into putting out a perfect job, without one whit of help from the technology of an automated zombie of a stove.

Slow Cooking

Probably the greatest glory of the wood range is its ability to top cook more slowly—more gently, really—than any other device known. While the soapstone griddle is heating over the firebox, there's a cast-iron skillet set to one side with a rasher of thick-sliced bacon strips gently rendering. By the time the griddle cakes are done we have bacon that is cooked through, crisp and browned on the outside, but still full-sized, or almost. Not all fried to a crisp with all fat and moisture driven out. Same's true of many other dishes, toast, for example. We just set a row of home-baked breads sliced thick on the hot-but-not-real-hot section of the cook top for two-to-five-minutes per side. The gentle heat toasts the outsides to a golden brown so it is wonderfully crisp, while the inside remains soft and moist. Nothing like toaster-done toast that's either raw or dried out. Louise puts slabs of butter on the first-toasted side, and lets it melt in while the other side toasts. Friend, you haven't tasted toast till you've had it straight from the top of a wood-burning range. (Though any stove will do a passable job, only a big range offers the wide variety of heats for complete flexibility in slow top cooking.)

Slow cooking of stews and the like is enjoying a new popularity—powered by electricity. You've seen ads for the newly popular "Crockpots," insulated cannisters containing a low-power electric heating coil. You fill the pot with stew parts, turn on a timer in the morning and that evening it is done. But that isn't genuine slow cooking as you can do it on a wood range. The back of the cook top is hot eight-to-nine months a year—boiling hot during a biscuit bake, but usually just warm enough to keep a pot or tea kettle barely bubbling. Not hot enough to make the bottom of a stew stick or boil the fluid off, just a gentle slow simmer. There's nothing like it from any modern appliance.

The stockpot sits at the back of the range. The same brew for a week at a time—maybe longer. Into it go all the meat scraps but smoked products and fish. From time to time melted fat is ladled off to be saved for soap. And whenever Louise needs a quick soup, she adds cut vegetables and barley to several cups of stock and boils till done. The stock is also the base for gravies and stews.

Slow cooking—simmering a bit faster than the stockpot, perhaps so that a pair of bubbles pop up every second or so—is *the* way to cook up an old rooster or a chunk of that grass-fed beef that didn't turn out quite as tender as we'd hoped. And, words can't describe what the low, slow and mellow heat of a medium-fired cook top can do to a brace of pork or lamb chops, lightly breaded by dipping in whipped egg, then in seasoned bread or cracker crumbs. Cut enough fat from the rinds of the

chop to rend into a thin layer of melted fat. Put the chops on, jiggle them back and forth a few times to assure they don't stick, then let them sit there for 30 minutes to an hour per side—just barely popping. Turn when the first side is golden brown and repeat. You'll never fast cook a chop again. These are moist and tender throughout, cooked through if you want (you must with pork) or with lamb or veal chops, pink in the middle. But all are unmatchable. You can have the same effect with any grilled food—potatoes, breaded zucchini, chicken or fish—slow frying or grilling over gentle wood heat is a kind way to cook food, and the food responds in kind. Just now, Louise is putting the final touches on a pot of slow-cooked ham and beans that has been back and forth to the range since noon, day before yesterday. Here's her recipe, and I guarantee no digestive problems: you can swallow the beans whole.

Louise's Slow-Cooked Ham and Beans

When the hambone is showing through along most of its length, take a pound of dried beans, sort and wash well. (We grow our own beans in with the sweet corn; last year the variety was Missouri Wonder, a fat, brown mottled bean from the Midwest.) Put the beans and three quarts of water in a sheet metal pot. (Cast iron isn't good for letting beans soak; it tempts rust.) Add a whole bay leaf, as many whole peppercorns as you like and salt—a pile in the palm of your hand the size of a quarter. Put the pot on the hottest part of the cook top, bring to a hard boil, then cover and push back to simmer for a quarter of an hour. This is to start the softening and to sterilize everything so the beans won't try to ferment and turn into bean beer.

Set the pot in a chilly place to soak for overnight at least. A day and night is better. Come morning of the day you want ham and beans for supper, bring the pot to a boil with the morning fire and keep it just barely simmering till midday, for five-to-six hours. Then cut up in bean-sized chunks and add: one big onion, two carrots, both peeled, and a stick of celery, top and all. Top up the water, taste and add salt if needed. If you like a tomato-y soup, put in a quart or two of tomatoes home-preserved in a thick tomato broth or a big can of tomato paste—more if it suits. If not all like the tomatoes (some of us do and some don't), those who do may add in as much of your good, home-done ketchup as they like at supper. Oh yes, put in the hambone, meat on, and most of the fat cut off.

Now, bring the pot to a soft boil again, then put it on the back of the stove and keep it so that a steady but very gentle flow of bubbles come up. You'll have to check frequently if baking to make sure it doesn't get so hot as to boil and stick. In another five-to-six hours of slow cooking, the beans will be soft, tender, and perfectly delicious. Before serving, boil the broth down as far as you want. Louise keeps it a fairly thin soup; the leftovers are grand for lunch.

Maw Wilson's Corn Cakes

With this you simply must serve corn cakes and buttermilk. You can find your own buttermilk, but here is how to make corn cakes—sort of a cross between corn bread, hush puppies and griddle cakes—a concoction out of south Missouri via Louise's mother, good Ol' Maw Wilson. For four moderate eaters, with a few cakes left over to toast for breakfast:

Heat the soapstone griddle to low griddle cake temperature, so water just skitters off. Then combine about a cup of stone-ground yellow cornmeal with a quarter cup of unbleached white flour and a good shake of salt. (You can vary the cornmeal/white flour proportions to suit yourself. Maw Wilson relies on pure corn.) Add in an egg—two if they are small, two tablespoooons or more melted butter, ham fat from the bean pot or any other liquid oil and a heaping teaspoon of baking *powder,* not the soda. Now stir in enough whole milk to make a griddlecake-consistency batter; a spoonful poured on the griddle doesn't sit in a lump, nor does it run all over, but it spreads out into a nice pancake shape. They cook just like pancakes. These proportions aren't exact since no one in Louise's family has ever measured ingredients exactly and when I ask what they mean by a "good shake," say of baking powder, the reply is: "Well, a good shake, but not a *real* good one, you understand. See?" And they hold out a hand with a mound of baking powder in it. "Just about this much." And that's cooking with wood.

By the way, that meal of corn and beans and enough meat to flavor it, perhaps with a bit of greens on the side, provides perfectly balanced nutrition—the corn/bean combination of complementary proteins is better than either ingredient alone, and you could live on it. Indeed, it came to south Missouri with the first settlers who felt themselves lucky indeed to be able to live on it. Of course, the European settlers got the goods from the Indians; all the corn and many types of beans are native to the Americas. Corn and beans are still the staple foods of some reservation Indians and a lot of other folks from our own Southwest clear into Central and South America. So, savor a bit of real all-American history and cook up a meal of beans and corn cakes on the wood stove. And it may not be all history. Some experts are predicting times as hard as pioneer days ahead as the world runs short of everything you can think of except hungry people. But those of us with a few acres cleared, and a good store of corn and bean seed, our iron pots, a stove and the wood to cook with will sail through. A bit slimmer than now perhaps, but well fed and warm. Good cookin!

The Range as Space Heater

It was a cold February when we moved in off the farm to the town place with the old Glenwood in the kitchen. Our second week the central heating went on the blink over a long holiday.

All we had to keep us from freezing was the range, and she did. The upstairs was a bit chilly around the edges, but that's what blankets are for. So when you read some expert telling you that cast-iron ranges are poor heaters because the fuel box is tiny and only takes twelve-to-sixteen-inch logs, you can bet he's never used one. The range is a tremendous heater, simply because of that great mass of cast iron soaking up and radiating the heat. Agreed, the fuel box is small compared with a big airtight heating stove. But that is so you can regulate your heat; the small fire cools or heats in a matter of minutes—essential flexibility for the cook. But, fed correctly, it heats—the Glenwood kept us warm all by herself for three cold days.

From the morning breakfast fire till the evening dish towels are hung on the drying rods beside the firebox, the range has been heating the kitchen and adjacent rooms, though her main job was to cook. Now it's time to bank the fire to throw heat during the night. There is always a good bed of coals in the firebox; I stir these well and, if the ash level is high, I may take a few cranks with the shaker to remove dead ash. I crank till a good "star shower" of small coals falls into the ash drawer.

Then the firebox is filled, with the biggest unsplit logs I can

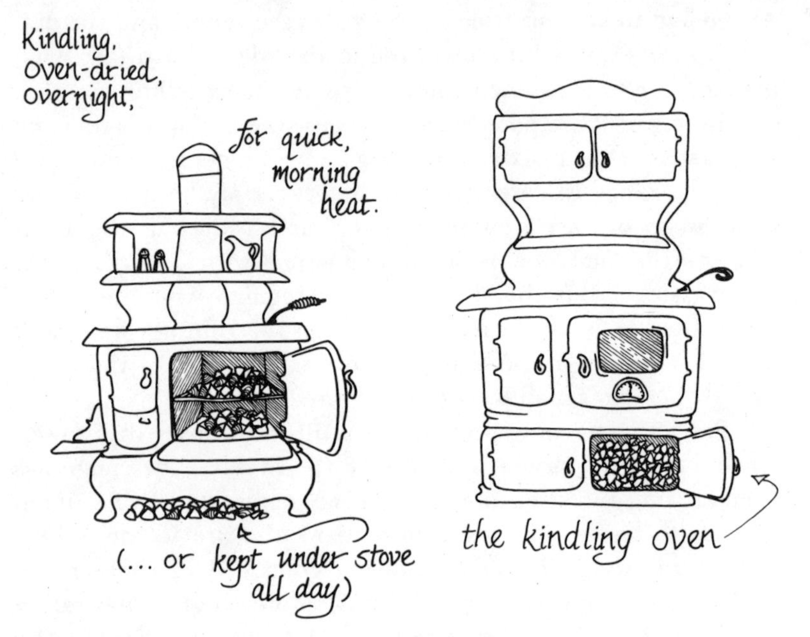

get in if the coals are hot, with quarter-splits on the bottom and round logs on top if coals are low. We open damper and draft till a good fire is caught, then all is closed down for the night. Usually one of us is up once or more during the night with one or another child, or a sudden urgent need for another cup of the bean soup that stays hot beside the stockpot at the back of the stove. The box is refilled if needed. This way the range heats like any good airtight. The kitchen never has a chill in the morning and is comfortably warm when we come down except on very cold, windy nights. Even then there is enough of a coal bed to get the fire going again for the morning coffee. The teapot is already hot enough to brew up a first cup. This is the best time to empty the ashes if needed, we shake down till "stars" fall into the drawer. A handful of splinters goes on, followed by a layer of small splits, not quite touching, and three one-quarter splits on top: Louise's magic arrangement once again. Draft and damper are opened and in 10 minutes the coffee water is boiling.

If the house is unusually cold we'll throw the oven control to let hot air circulate through the whole range for a half hour or so. This requires more draft, more wood for a hotter fire with the result that a great deal more heat is radiated into the air. Our oven door is always left open at night so the fire side of the oven

will contribute the heat it absorbs. But the firebox alone is a good enough night-long heater—and on our range much more economical than when the oven is heated also.

So, in addition to cooking your meals, the range will heat a room or two easily and keep the whole house comfortable in an emergency. We figure that we use about one-tenth of a cord (five dollars a week at most) in the range during a cold week for all cooking and perhaps a third of our heat. For heat alone, and using quarter-splits, we average one stick an hour for mild heat, one per half hour for medium heat and two per half hour to toast ourselves after an afternoon of tobogganing down the back hill.

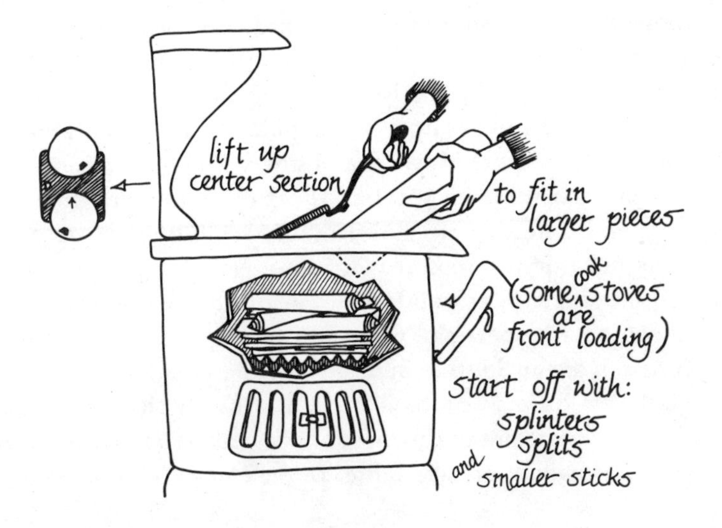

CHAPTER SIX

Getting the Wood In

Now that your fireplace or stove is in and operable, you'll need some wood. It's easiest is to buy it from a woodlot operator who has the equipment to turn out cordwood on a commercially profitable scale. Look for ads in the newspapers of the nearest rural towns. This is not to say that the fellow who puts up a cord or two on weekends should be avoided; likely he will give you a better price than the full-time woodman, but you'll have to ask around to find him. Besides, a man who has the tractors and logsplitter and all has a considerable stake in getting and keeping customers. By and large, if he tells you his wood is good, it will be, and he won't try to pass off newly cut wood for well aged. He'll probably be particularly friendly if you let him know you will be in the market for a half-dozen or more cords a season. But you should still know what you are (or should be) buying.

Selecting the Best-Heating Woods

First, if at all possible, insist on hardwood. Most coniferous woods such as white pine are lightweight and resinous. They will burn, in fact, they make the best kindling there is. But they last no time at all and can get up a roaring blaze that not only wastes wood but can very quickly overheat a stove. Too, the resinous sap will not burn completely, unless the fire is too hot for best economy and safety. The unburnt sap will deposit on your stovepipe and flue as creosote. Old-time New Englanders never burned pine and similar woods out of justifiable fear of flue fires.

Of course, the species of wood you get depends on where you live. In much of the South, you'll be burning pine because that is about all you have available. Same is true in the coniferious forests of the West, but in the most heavily populated areas of the cold belt, hardwoods are available. In the chart, "Characteristics of Several Wood Varieties," we've listed comparative weight, splitability and such for the most common woods in the East and Midwest. Here in our part of New England, rock maple, birches and hickory predominate. Farther North, the woods turn to conifers, down South there are more beech and oaks and the hickory gives way to black walnut. All burn well. One tip if you are new to wood and wood heat; look for firewood in the highlands. Bottomland trees run to cottonwood, sycamore and soft (red) maple that are watery and take a long time to dry. Some aren't worth much as fuel even then. Don't try to find any chestnut, though most of the wood heat books around dutifully list its characteristics along with other fuelwood data gleaned from out-of-date foresty publications. There hasn't been any American chestnut around since the chestnut blight cleaned it out several generations back.

If you want more information on trees, including the best identification guide to standing timber (photos of the tree in leaf and out, bark, leaves and seeds and more), send for the book *Knowing Your Trees* by G. H. Collingwood and Warren D. Brush from the American Forestry Association, 919 Seventeenth Street, N.W., Washington, DC 20006. The price is probably around $10 these days but the book is the best there is.

Buying Firewood

As we must have mentioned earlier, wood is sold by the standard cord, a pile of four-foot-long logs piled four-feet high in a stack eight-feet long. When you buy it, big logs should be split and cut to whatever length you need. Four-by-four-by-eight feet or a pile 128 cubic feet in volume no matter how you cut it, though. Now, dried wood weighs up to 63 pounds per cubic foot (for shagbark hickory, the most dense of common hardwoods). About a third of the volume of any cord is airspace and most woods are lighter than hickory, so your typical cord of well-dried hardwood should run around two tons in weight. Obviously, if someone tries to tell you his little half-ton pickup contains a cord of wood, he's gotta be kidding. Don't let him kid you.

Most professional wood sellers will have some sort of dump body on their delivery truck. I've seen one- and two-cord loads delivered by a big five-ton dump truck and half-cords delivered on a beefed up light truck with a makeshift dump bed on the

Sam, Martha and their pal Christie are stirring up an honest half-cord of assorted hardwood limb pieces, cut to 12-inch stove length. Note the deep tire tracks on the lower left. This is early spring, mud time in New England, and the truck got bogged down and couldn't get back to the wood porch. The kids hauled the bulk of it to the porch, stick by stick, over the next few days and got a kick out of really helping. They keep the kitchen woodbox full of kindling, too. Well, usually they do.

back. Some years back, Louise and I bought a cord of wood as stacked and transported it home in the back of a Volkswagen. It took quite a few trips. I'd advise you to have your wood delivered even if it does cost an extra $10 a cord (as seems to be the case hereabouts these days). Remember, the heating value of a cord of well-aged hardwood is the equivalent of 200 gallons of No. 2 fuel oil, about $80 at today's rates. If you spend $40 a cord for the wood plus another $10 for delivery—even another $5 or so to have it stacked, you are ahead of the game.

Actually, don't have it stacked or split into small sticks. Do the stacking and splitting and hauling yourself. It is easy work; no one stick weighs much. But piling up several tons of wood one or two sticks at a time is perfect exercise, particularly if, like most Americans your regular work is sedentary. Splitting up and hauling in a hundred pounds of wood every evening gets your heart pumping and blood moving at a moderate pace for five or ten minutes every day. Then, there's the bending and shovelling and hauling of the ashes, most likely into the cellar several times a week. Honestly, keeping a couple of wood stoves fed and cleaned will supply about all the exercise you need to keep reasonably fit during the winter. Spring cleaning entails a lot of elbow grease too. Any wood-heated house picks up a covering of the gentle rain of ash that escapes every stove or fireplace I've ever seen at each loading and clean out. Not "dirty dirt" as my Grandmother used to say, but still requiring a good scrubbing now and again. No doubt about it, wood heat is hard work from start to finish. But if you take it up, then garden summers like the fellow whose old stove we refurbished a while back, you may live as long as he did. He died prematurely at 99.

Reconnoitering

If you've reason to suspect the reliability of your wood seller, it's best to visit his yard before you buy. Chances are that he won't have wood sitting around in neat cords but stacked here and there in piles. An experienced wood stacker can tell when he has a cord or half-cord on his truck by the look of the pile or the settling of the springs. Besides, your cord and mine will never contain the same amount of wood. Back when charcoal burners bought cords of wood where and as stacked, the loggers worked out some pretty dandy ways of building the maximum amount of air into a cord. Take the seller's word, restack back home if you like and complain if you feel you've been shortsticked.

Drying Firewood

Do check the wood for dryness, though. Even fully air-dry wood will contain 20 to 25 percent moisture. However, some green woods are over one-third water by weight. You want as little of this moisture as possible in your fire. Green wood is hard to light. Too much heat energy is lost turning the moisture into steam and driving it out the flue. Check the cut ends of a sampling of the wood. Green wood will show the growth rings and perhaps a surface roughness from the saw, but the cut face will be solid. As wood dries, it shrinks in circumference—never lengthwise. Shrinking takes place unevenly and stresses occur in

Here is a stack for well-dried cordwood. Note the cracks that indicate dryness (the wood is six months off the stump) are different sizes and come in different patterns depending on tree species and age.

And here, for comparison, are two sticks off the same small (quince) tree. The year-old branch is dark colored and well-split in the characteristic ray pattern. The day-old stick shows the beginning of dry-cracks in the center. The cracks will expand out ever so slowly as the wood dries.

the wood, resulting in small cracks. Most species show cracks running from the center to the bark. Others will crack along the annual ring. Still others will show random checking. With a bit of experience you will also be able to distinquish the dull thud of a wet stick being knocked against another from the slightly crisper sound of dry wood.

The air-drying process is a function of time, taking place faster in warm dry seasons than wet or cold. Basically, the wood has to lose its inner moisture to the less saturated air. Split wood dries half again as fast as whole logs, some species faster than others. If the wood has been cut and split, then aged over summer it can be as young as four months and burn well. If piled in four-foot unsplit lengths in the woods, then cut and split, it should be the better part of a year old.

Sitting out in rain or snow has little effect on the wood's essential dryness, except for the top and ground layers, which may have wet bark. Any surface moisture will evaporate overnight so it's a good idea to get all of tomorrow's wood in this evening. Stack kindling in the kitchen oven or under the range.

Mixed Cords

Now, you shouldn't expect a cordwood seller to bring cords of only the best-burning species. His woodlot will contain a mixture of species, and he'll cut and sell them all. If his cords appear to consist mainly of the low-weight kinds such as white birch and black cherry, offer him one-quarter less than he asks, particularly if you know another dealer who has a mixture with more maple, oak, or other dense wood.

Especially beware of a seller offering you cords of American elm. The wood burns beautifully, don't get me wrong. But it is lightweight when dry—contains over one-third moisture by weight, when green. Also, it is almost impossible to split. Not only is the grain all twisted around, it is a highly elastic and springy wood. Every old-timer out our way has his story of the greenhorn sent out to split an elm log who was last seen by human eyes, whacking away as it bounced along into the woods ahead of him. Further, elm may be acting as a nursery to bark beetles, a million or so of them to each recently dead tree. If you store the wood inside, you may one day find yourself godparent to a whole lot of harmless, but bothersome little bugs. Actually, if elm has been aged for a proper length of time, any bugs will be long gone. It's the same for other species. The chances of importing a nest of termites or carpenter ants, for example, along with the woodpile is nil. You'd have to be unlucky enough to get the brood chamber and fertile queen, which in most species are well underground, not in the wood at all. The only pests we've ever had in the woodpile were a nest of bumblebees and innumerable field mouse nests, but they all moved in well after the wood was stacked. I wouldn't worry about bringing any harmful critters from the woods into your woodshed.

CHARACTERISTICS OF SEVERAL WOOD VARIETIES

Common Northern Hardwoods	Weight of one cubic foot, dry	Fuel Efficiency Ranking	How Does It Split	Wet, Medium or Dry
Hickory, Shagbark	63 lbs.	1	well	M
Hornbeam (Ironwood)	49 lbs.	2	hard	D
Birch, Black	48 lbs.	3	fair	W
Beech	45 lbs.	4	hard	D
Maple, Sugar	44 lbs.	5	fair	M
Oak, Black	43 lbs.	6	fair	W
Ash, White	42 lbs.	7	well	D
Birch, Yellow	40 lbs.	8	hard	M
Maple, Red	38 lbs.	9	fair	W
Birch, White (paper)	37 lbs.	10	easy	W
Cherry, Black	36 lbs.	11	fair	D
Sycamore	35 lbs.	12	doesn't	W
Elm, American (white)	34 lbs.	13	doesn't	W
Two Coniferous Softwoods				
Spruce, Red	28 lbs.	14	easy	D
Pine, White	25 lbs.	15	v. easy	D

Storing Firewood

Speaking of a woodshed, if you have one, great. In our various homes we've stacked wood in a genuine woodshed, a slatted corncrib, the cellar, under a big porch and, just now, on a two-cord-sized back porch. Store your wood where you have the space. The major consideration is to minimize time and distance in hauling. Ideally, you'll be able to cut wood into stove or fireplace lengths in the woods, toss them directly into a truck then restack in a shed just a few steps from the back door to dry under cover for a year. That requires a lot of stacking room that we don't have anymore.

We cut wood into standard four-foot cord lengths and leave it in ricks in the woods over winter, then cut it to size and haul it in several times over the fall and winter. The back porch is just outside the kitchen and bringing in the wood is a short walk (that the kids enjoy most evenings). I'd advise you to work out the easiest, least time-consuming system of bringing in the wood that you can before you take on the first cord. If you simply have to store it at a distance, get a big garden cart (the biggest and best I know of comes from Garden Way Research, Charlotte, VT

05445) and pull several day's requirements up to a convenient door or window every so often. Having to hand-carry armfuls over long distances several times a day will add too much labor and, in time, boredom, to your wood heat experience. And you don't *have* to store all your wood under cover. Dry wood contains 25 percent moisture, about the average humidity in the air. Even in rain or snow, wood in the heart of your pile won't absorb more water than that. So, you can just stack it out beside the house if you like. Bring in a two-day supply to dry off for a day or more and you'll have good fires.

Arranging the Woodpile

It may seem like a minor detail, but when you are stacking your wood, it's a good idea to divide it into "grades," particularly if you have an unlighted and unheated woodshed and, come winter, will want to get out and back with the type wood you need, and quick. I try to separate wood into four categories, either in separate piles or separate sections of the pile.

No. 1 is the quick-burning stuff in order: splinters, small kindling splits of anything and quarter-splits of light woods such

Louise brings in the first sticks from a newly stacked cord of mixed hardwoods. There are three faces of foot-long logs at the rear, a tenth of a cord of quarter-splits (a week's supply for the range) to the left, and, in the box, several day's requirements of small stuff for kindling and biscuit wood.

as the birches. No. 2 is the same woods, but in half-splits or small logs. No. 3 is composed of whole round logs of the light woods plus quarter- or half-splits of harder woods. The fourth, and really most important from a wood-garnering-at-midnight point of view are whole logs (any size, but the bigger the better) of the really dense, long-burning woods. When the fire gets low at 4:00 A.M. of a -30 degree F. winter evening, you want to get out to the woodshed and back quick, and with the most firepower you can grab. So in pile No. 4 go the hickory, hornbeam, oak and other such logs that will burn for a good four hours. Stack it so you can get it blindfolded and when you have to, you can.

Tips for City Folks

If you live well into a metropolitan complex, away from the woodlots, you'll likely run into one or another of the tricks some of your crafty country cousins like to pull with their wood. One is to charge you double or triple the price a cord would bring a half hour drive out of town. The winter of 1974/75, first year of the "energy crunch," wood was bringing $100 a cord in Boston. Maybe worth it if you have more money than time to spend and only need a few logs for their entertainment value. But if you are serious about heating with wood, it will pay you to get out into the woodlands and scout around for good wood at a fair price. I'd say, spend one weekend scouting, and if you can't find a wood dealer who will deliver for a reasonable fee, then the next week rent a really big truck (one of those moving vans would do though a dump truck is best) and haul the wood in yourself. Say you need five cords, the truck costs $45 and it takes you a good eight-hour day to do the hauling and stacking. The difference between $40 and $100 cords amounts to $255, van costs deducted. Pay yourself more than $30 an hour.

Be sure to get *well* out of town. Louise and I did a wood survey in the Boston area not long ago. Right in town there were trucks loaded with puny little elm and white birch limbs, being sold by the half-bushel basket full: two dollars for the elm, three for the birch. The sellers were hawking "cords" of wood. A little way out of town, in the high-income Concord/Lexington area a "cord" consisted of a wooden frame about two-feet square filled with fair-sized limbs. Five dollars for that. Even farther out, but still in the suburbs, a fellow was selling what I'd call a "face cord," though it was an honest cord to him: a stack four-feet high and eight-feet long, but only a foot deep. A quarter-cord, in fact. (A face cord is any four-by-eight foot stack of logs cut to burning length; could be any depth.) This man wanted $25 for his quarter-cord, and the buyer had to haul it off.

You may just run into someone who insists that his cord, three feet by five feet by six feet is a legal cord. He's legally correct; 90 cubic feet of wood is a cord according to people who measure heating values. It must mean 90 cubic feet of solid wood, though. The difference between that and the 128 cubic feet in a conventional fuelwood cord represents the air. If anyone tries to sell you a legal cord at regular cord price, $40 to $50 this year, offer him 70 percent. That would be $28 on the ground, $35 delivered.

Cutting Your Own

There are plenty of places you can look to get your own wood, free for the hauling, or perhaps for a small fee. More and more federal and state parks are being opened for supervised cutting. I'd say, scout around for such opportunities, perhaps by checking with the state forester (at the extension service of your state's land-grant college). There's at least one in every state. Vermont has a forester in each county, or so I'm told.

Any heavily used government-owned wooded areas, including roadside picnic areas and parks in town will be frequently culled of dead trees and limbs, if only to remove a potential hazard to visitors. Check with the street and park department. If they won't let you cut down trees, you may be able to pick up windfalls in the spring. Any area where American elms grow (grew) will have an elm removal unit. Most are professional landscapers contracted by the city. They come in with heavy equipment and take out a big tree in half a day. You may not be able to pitch in, but you can find out where they take the elms and any other trees removed. Most places, the elms are supposed to be buried or burned, and you should be able to get permission to take out as much of the smaller wood as you can.

Sources

An excellent source for one of the best-burning woods there is is any nearby fruit orchard. Most fruit trees are pruned annually, and a good orchard keeper is constantly culling diseased or too-old trees and planting new stock. Most will let you take out the old wood and if you do a good neat job, they may even pay you for doing it.

Nurseries, landscapers and the local "tree expert" are all possible wood sources. So are lumberyards, where shavings or sawdust may be available in enough quantity to justify your making a sawdust stove. In logging country (and you'd be surprised how much small-scale logging activity exists near population centers) you can find good fuelwood at the sawmills or out in the woods. Loggers or landowners are usually only too happy to let you clean up the tops of trees cut for lumber. Sawmills will have huge sawdust piles and if it is a mill that processes logs fresh from the woods, there will be great stacks of slabs, the bark and outer wood removed to square up the logs. In our area, the slabs are strapped up in approximately cord-sized bales and sell for $10 to $12 each. You get a lot of bark and air, but it burns and it is cheap. Be sure to get hardwood slabs. Pine slabs aren't worth hauling home.

Any factory or mill that works with wood will have scrap. Foolishly, in my opinion, clean air ordinances may keep them from burning it themselves for heat and you may be able to get chunks of beautifully aged cabinet or other milled wood, possibly free for the hauling.

Buying a Woodlot

Finally, you may want to buy standing timber or a woodlot. Landowners often sell timber on the stump for $100 or $200 an acre; this would be for all the timber of marketable size. The owner will expect you to leave the small stuff to grow up and net him a few dollars in a few more years. Perhaps you can get permission to cull out only dead or diseased trees for the wood, and if you pay for the privilege—or if you pay very much—you are getting the worst end of the bargain. Purchasing the land itself purely for the wood may make sense in a lot of places where unimproved land is cheap and plentiful. Make sure that land borders a public road or you have *deeded vehicular access,* a right-of-way in. A plot of 12 acres or more, well treed with a variety of hardwoods, should turn out all the firewood you can use at an average rate of one cord per acre per year—and it will keep on producing indefinitely. Be sure to make an informal timber cruise of the land before you spend any money. Go over it all in parallel paths, so you see every tree on it. Of course, you want a majority of dense, good-burning hardwoods. If there is a lot of big timber, you'll have a hard time getting it cut into firewood. It's best to find a stand of young trees (as we do in cutting our own), and plan to leave the big logs or have them taken out by professional loggers with the necessary equipment.

In cutting your own, you'll need a chain saw and some other gear mentioned in the next section. I'd say, rent it if you are only spending a few days a year in the woods, getting in a cord or two. Good saws cost $200 to $350 and they need a lot of maintenance. It's best to let the rental agent do it unless you are really into

(what little boy couldn't spend hours in the fragrant sawdust under his father's sawbuck?)

wood on a significant scale. A couple of good men with rented saws and a borrowed or rented pickup can get a whole lot of wood out of a park or an orchard in two or three weekends.

In most any kind of gleaning and in working your own woodlot, there will be a lot of little branches you shouldn't waste. The small stuff is usually flexible and willowy enough that a chain saw just wiggles it around, without cutting it. Probably the easiest thing to do is cut it into handy four-foot or so lengths and store it that way till needed. Then to cut it up, build yourself a sawbuck, traditionally a pair of **X**-shaped ends attached to each other with two or three longitudinal stringers about a yard apart. Ours, for cutting 12-inch stove wood has three **X**s, one just under one foot from one end. That way I can cut any length log we need. For sawing up the little stuff, spend a few dollars for a 30-inch Swedish bow saw. Blades are ultrathin and sharp and cheap to replace. There is an antique bucksaw and a two-man crosscut out on the farm, but both take wide cuts and are heavy as can be. When woodcutting time comes around, it's not so bad living in the mechanized 1970s after all.

Bow Saw
(Large)
Smaller,
lighter,
Swedish Bow Saw
two-man
crosscut
twisting...
... tightens
the blade
Bucksaw

Managing the Woodlot Naturally

Most books and government publications on woodlot management that I've seen tell you to grow the maximum number of marketable sawlogs. A mill will pay from $10 for a mature white birch to several hundred for an especially fine black walnut log. There's a whole profession, forestry, geared to thinning, pruning, and selective cutting for sawlog production. And the industry set up to use sawlogs and forest products rivals the oil companies in power to work up depletion allowances, special tax breaks and freight rates. (It's cheaper to ship new pulpwood for paper than to ship old paper for recycling.) Well, foresters are good men, outdoorsy and sincere in their work. And if I'm let rant on about the abuses of the industry you'll get bored and I'll likely get apoplexy.

Retaining the Natural Mix

We and our woodcutting friends feel that the woods ought to be let grow as naturally as possible. Every tree has a purpose, whether we understand it or not. We intend to retain the natural mix of hardwoods, coniferous trees, scrub and meadows that nature left us. It just seems that in the long run, thinning a forest to the few species and tree shapes and sizes that make the most profit for the industry is as unnatural as growing nothing but corn on your farmland year after year. Monoculture is monoculture, and with trees as with food plants, it is asking for epidemic disease, soil imbalance and an upset ecology that will ultimately call for corrective measures such as the sprays they are using to control gypsy moths in the artificial in-town plantings of the East and the several recent insect epidemics in the single-species lumberwood plantings of Douglas fir and the like out West. Nope, we let nature mix up the trees as she will. I guess we're organic foresters.

Our woodlot has been logged three times, twice since the turn of the century, and still it bounces right back with enough to spare that we can cut our fuelwood supply and never make a dent. And that's the way we'll keep it—we and the trees and all else that lives there, existing in natural harmony, we letting nature make the big decisions and nature letting us remove enough wood to keep us warm. The oddly tall maple across the road will remain: Baltimore orioles nest in the highest branches every spring. The sunny, but scrub-choked glades in the woods will stay as they are: mice nibble the lower few inches, rabbits feed in the next foot and few inches, and deer browse the upper layer.

I know for a fact that rabbits use the humps and hollows of a huge "unthrifty" rogue white pine's roots for winter quarters. A whole succession of them some winters, as the live traps turn up a good rabbit stew from the tree once or twice a month in snow time. I'm almost as sure that the same quarters are home to litters of one or another critter each spring. That's what the dogs say, snuffling around the needles each time we pass the tree on a walk.

By the rules of modern forestry, this old tree would be one of the first to go. The term rogue applied to timber means a tree that sprouts lateral branches low down on the trunk, so the sawlog is too short. New England is just now growing back from the deforestation that commenced in colonial times and continued till the rocky hillsides became uneconomical to farm in the 1930s and 1940s. Even in the early days they considered "rogues" too poor to cut. But they didn't cull them. They left them grow to shelter wildlife and act as seed trees.

You find one of these out-sized, odd-shaped giants every 100 yards or so in the woods. The dogleg along our western boundary was marked by a huge chestnut till the blight struck. The stump's still there, sending up shoots that are blighted after a year or two. And there is a fantastic lightning-split black oak where the deer paths meet up at the north corner. The tallest American elm I have ever seen is down by the road. We'd hoped that it might be one of the few to hold out against the Dutch elm disease, but last year the leaves browned early up in the crown. Bark is sloughing off the upper limbs now, and Louise says the highway department has affixed one of its metal tags scheduling the tree for removal. Barring such disease, fire or storm, though, all of the big trees will remain. They are what make our woods a forest. It's their sons and daughters that we use for fuel.

A Cord an Acre a Year

Now, the old-time rule of thumb is that an acre of land will produce about a cord of wood a year indefinitely. But don't plan

A rogue tree
"cabbage pine"

to bull your truck or tractor over your whole woodlot each year to extract one cord from each odd acre or so. Most of our own woods is left alone; fuelwood is cut from several scattered stands of the right-sized trees. For sheer ease of handling I prefer them "half-cord" sized, with bases of perhaps a foot to 18 inches in diameter. The trunk of a tree that size will make up about a quarter-cord when cut and split, and the limbs and larger branches will fill out the rest of the half-cord. A few trees such as white birch seldom grow much over good cordwood size whereas a maple, oak or ash is a mere baby when still that small. Eight to twelve trees of the proper size will supply our year's wood needs.

Of course, our initial cutting on the woodlot was to clean out injured and diseased trees and gather up any fallen limbs that were recent enough to be unrotted and burnable. No reason to

waste good wood. We worked especially hard to remove diseased elms. In those days it was believed that the bark beetle that carries Dutch elm disease could fly only a short distance, and "tree-hopped" through the woods. Supposedly, if you cut down the infected trees, others nearby would be spared. Recent research shows that the little bugs can fly up to five miles looking for a healthy elm to feed on and a diseased one where they will lay eggs under the bark. So, from a prevention standpoint, culling sick elms may be a waste of time. However, the bleached, debarked skeleton of a long-dead elm is a saddening spectacle in the woods, and we are glad that ours are gone.

The Modest Clear Cut

In the process of getting to elms, split birches and the occasional maple showing the white, hard shelf-like fruiting members of a common decay fungus, we found we were tearing up the woods with the truck and felled trees. That's when we decided that taking that one cord per acre each year wasn't practical. Now that the larger culls are gone, we use a form of clear-cutting for the firewood supply. We choose a good stand of proper-sized trees and go right through it until it is cut out.

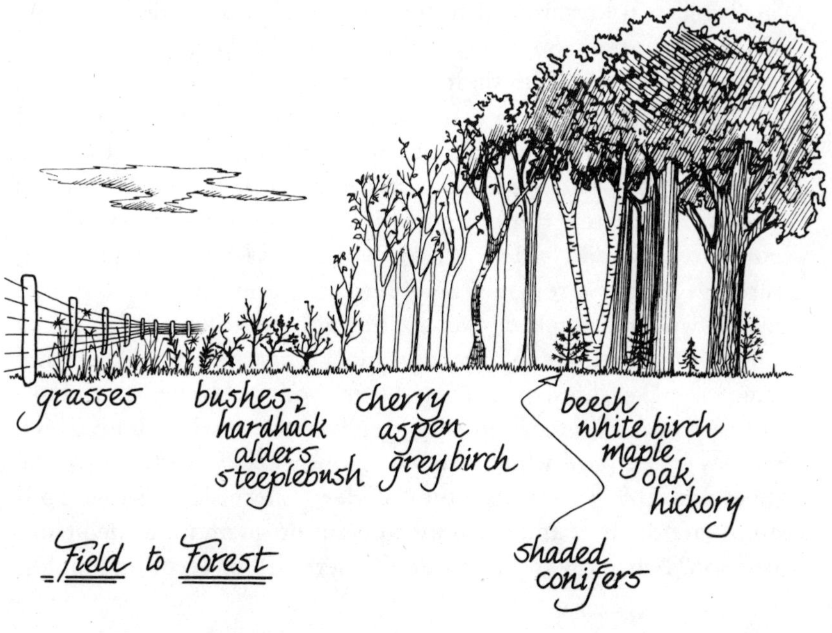

In the old days the tillable parts of New England were denuded of all but rogues, a few pasture oaks left to shade the cattle, and trees along roads, fencelines and walls. Then, as fields were abandoned gradually over the last century, each was seeded by the nearest mature trees. Typically, there would first be a stand of quick-growing scrub or "nurse trees" such as aspen, alder, grey birch or sumac to shade the tender, slow-growing seedlings of the next generation, maples perhaps, or other longer-lived hardwoods. And as the hardwoods grow they in turn shelter seedlings of the shade-tolerant conifers that (in the colder areas of the continent) will replace them as they fall or are cut in the natural course of time.

So, we find many stands of like-sized trees of the same species or several that commonly grow together. Maple/beech/birch is a common mixture in our area as is oak/maple/hickory. In the bottomlands we often find a few sycamores mixed in with grey birch and more common hardwoods. Many abandoned farm sites still contain aged apple trees (finest firewood of all). Throughout the woods are scattered stands of American elm, an occasional white or black ash, ironwood (hornbeam) and black cherry trees along with a rare basswood, persimmon, tupelo or sassafrass. When we find a white oak it is marked for preservation. Too many of the species have been cut and the wildlife so love the sweet white oak acorns that very few survive to propagate the tree. When and if we find an American chestnut or elm growing in apparent health, it is noted for perservation and continued attention. Someday a mutation will come along to preserve each species from their respective diseases.

However, there are a dozen or so stands of the right-sized trees on the lot and each year I cut a road into a new section of a stand and cut out our five or more cords of wood—details on logging a bit later. But our objective is to cut out all trees of half-cord-size age and older. There will always be a profusion of saplings ready to replace the older trees, and once the big ones are down and stacked, we go through, cutting out runty or unhealthy looking saplings, any black cherries with black knot fungus, oaks infested with galls or young birches that have cracked under an ice load. We aren't too particular about which species we cut and which are left. Nature has already set up the most desirable (from her point of view) succession. All we do is eliminate the halt and the lame and cut down on unneeded competition. Normally in these stands, there will be seedling maples, say, already 20 feet high, but only a few inches through at the base. With both the big trees and a lot of the runts gone, they will have free sky and soil to put on weight. Then when we come back in 15 or 20 years, they will be nicely grown to cordwood and will come down—or about three-quarters of them will, along with all of the other species, if we've decided to enlarge the sugar bush. In the interim we'll have a sunny glade to picnic in. The scrub will grow up for the deer to nibble and the neatly piled "slash"—small branches and twigs left over—will shelter the rabbits when the snow comes. Thus, we will repeatedly cut out selected patches here and there. Most of the woods will grow on undisturbed to become whatever type forest nature intends. If anything, the frequent little clearings and piles of drying wood offer a bit of variety to a walk in the woods.

Cutting Firewood

The traditional way of putting up cordwood is to cut a tree down, and into four-foot-long sections during the fall and winter, leave it to age over the coming summer, then put it into the woodshed the following fall. After trying to girdle trees and let

them age on the stump and also trying to cut trees during the slack time of midsummer, I've returned pretty much to the old way. The girdling (cutting a groove into the bark all around the base of the trunk) doesn't always work and in any event the girdled tree can take several years to die. Summer-cut trees are still in leaf, making it hard to see limb location so you can best judge the direction of fall. Leaves are a nuisance when you are trying to cut up limbs, too. They make stacking impossible, so you have to come back after they've wilted and dried. So, you may as well do the whole thing after leaf fall.

Equipment

The most essential piece of equipment in cutting wood is a wood mover. If you are hauling sawlogs out of rough country, the best thing is a team of draft horses. For cordwood, a farm tractor and wagon is good if you have one. If you don't, a good pickup truck will do for the wood and a lot of other hauling chores. Four-wheel drive is pretty essential, and "heavy-duty" suspension and cooling is good too.

A heavy-duty power winch attached to the frame, front or rear is good insurance if you get bogged in, but get a real (Warn

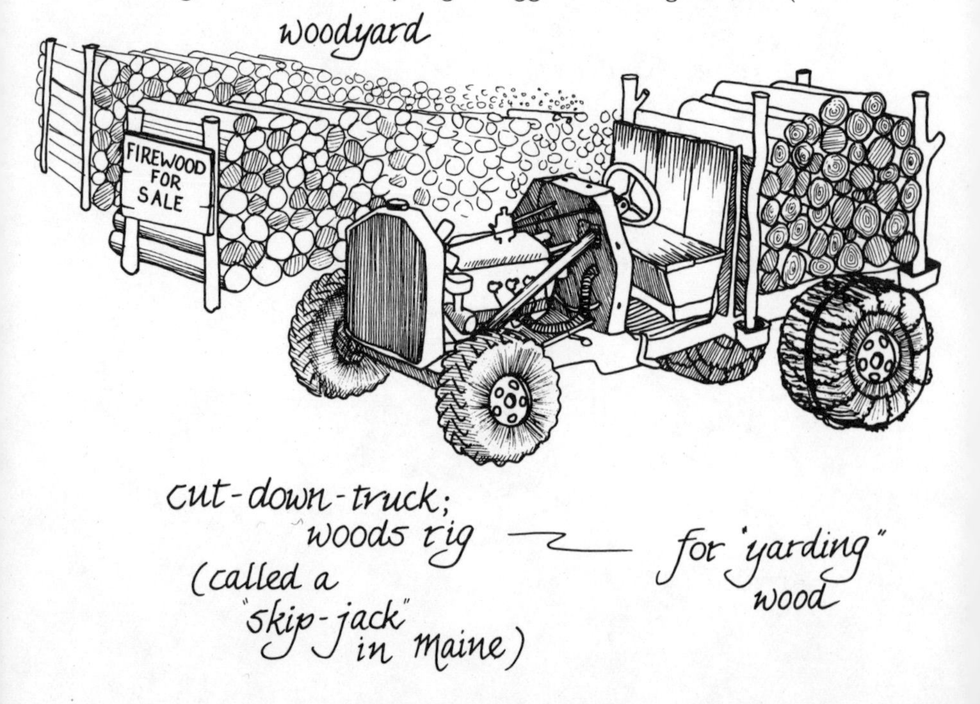

or other manufacturer) winch, not one of those little half-ton capacity toys you see advertised in the discount stores and mail-order catalogs. For rough-country travel, see if you can't have a gas tank protection plate installed too. All power equipment that goes into the woods must have spark-arrest mufflers to prevent fires—by law in most places.

Chain Saws

Unless you have the ambition to fell with an ax and cut up your trees with a hand saw, you'll need a chain saw. Once again, avoid the heavily advertised lightweight saws, gas or electric. If there's one piece of advice I feel strongest about, it is the saw. Get the largest, most expensive model you can swing. Make it Homelite, Remington, Stihl, Poulans or any other brand, U.S. or imported, that makes saws for professional loggers. That little "Mighty-Midget" gas or electric you see for just under $100 in the discount store may be OK for pruning the lilac bushes, but it won't last through a cord of wood. Roller tips are nice (we have them on the Remington) ; they let you keep a tighter chain but require added maintenance and a special grease gun, so the pros don't bother with them. Avoid saws with automatic features—chain oiling or chain sharpening gadgets; they're not necessary and are the first things that will break down on you.

Maintenance

When you buy the saw be sure to get three or four of the special round files for sharpening saw teeth and a small flat file for cutting down the steel nubbins—the spacers between each tooth—that kick sawdust out of the cut. The saw will come with a

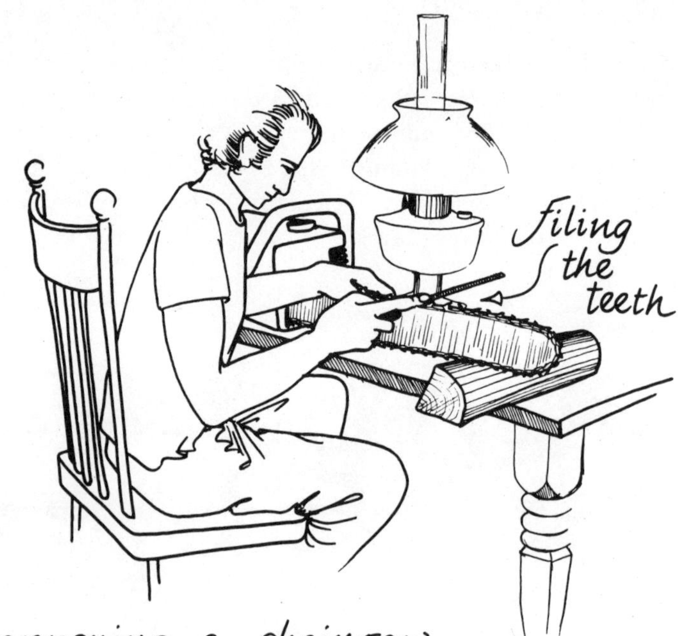

sharpening a chainsaw

comprehensive instruction manual including routine safety tips. All I can add is, be sure to learn and follow the advice. A live chain saw is considered a deadly weapon in the logging country and just one slip can leave you short a leg. The main rule is to hold the grips tight and never push on the cutting bar. Let the saw do the work at its own pace. If it stops cutting, makes smoke or requires more pressure than perhaps a gentle rocking motion your chain needs adjustment or sharpening. Any time you see burnt sap building up on the top back side of the teeth you should stop and go over the chain with the files. It is a good idea to give them a lick or two every 15 minutes anyway.

Saw noise is harsh, and prolonged exposure can damage your eardrums. Wherever you get your saw, you may be able to pick up some of the glass fiber stuffing professional loggers put in their ears; it only cuts out the harsh, painful and potentially dangerous wavelengths. Cotton isn't any good, but a set of earmuffs will do double duty when it's cold. Look for Flents earplugs in the drugstore.

It isn't a bad idea to buy a small brush and use it to keep too much sawdust from clogging the cooling fins on the cylinder in the woods or the saw may overheat. Then when the cutting season is over, you'll want to clean the saw up completely and store it. Use high-pressure air or water to clean out the fins completely. There will be a lot of oil and sawdust stuck on the engine and in various cracks and crannies and you may have to spray on gasoline (outside, away from any fire, of course) before spraying or air-cleaning it. Then polish up the outside, empty the gas tank, start it and let all the gas run out of the carburetor. If it isn't running perfectly, tune it, by the way. Instructions are in the manual.

Then remove the spark plug, put a few drops of light engine oil in the cylinder, give the starting rope a few pulls to coat the inside of the motor with oil, and replace the plug. Put a light coating of oil all over the saw, the blade in particular. Finally, pump out a good supply of chain lubricating oil, and move the chain around by hand till the bar groove is filled with oil. Empty the chain oiler reservoir, pump out the little that remains in the pump, and put the saw away till next use.

Felling

Although felling small cordwood trees isn't as dangerous as cutting down a yard-wide sawlog, any falling tree can kill you. You should know or decide where the tree is going to fall, and make sure it goes down there while you are a good tree-length away on the off side. Standing well away from the tree, hold a pencil at arms length, loosely so it dangles straight down. Compare with the tree trunk and you'll see which way it leans. Put the pencil tip at the midpoint of the base and you'll be able to see which side is overheavy with limb growth. Walk around the tree, sighting past the pencil and compare lean with weight distribution. You'll probably make correct judgments your first few trees because you put the time into looking them over. It's later, when you've more experience and aren't bothering with the pencil anymore, that you'll make mistakes. Or that is how it has been with me.

You will surely want the tree to fall into a cleared area. If it goes down into more growth, it will likely hang up in the branches of another tree, and a hung-up tree is a killer. The only thing worse is a "widow-maker," a big—often decayed—limb that a falling tree knocks off itself or a nearby tree; the limb will fall straight down, not ever over with the tree, and more than one

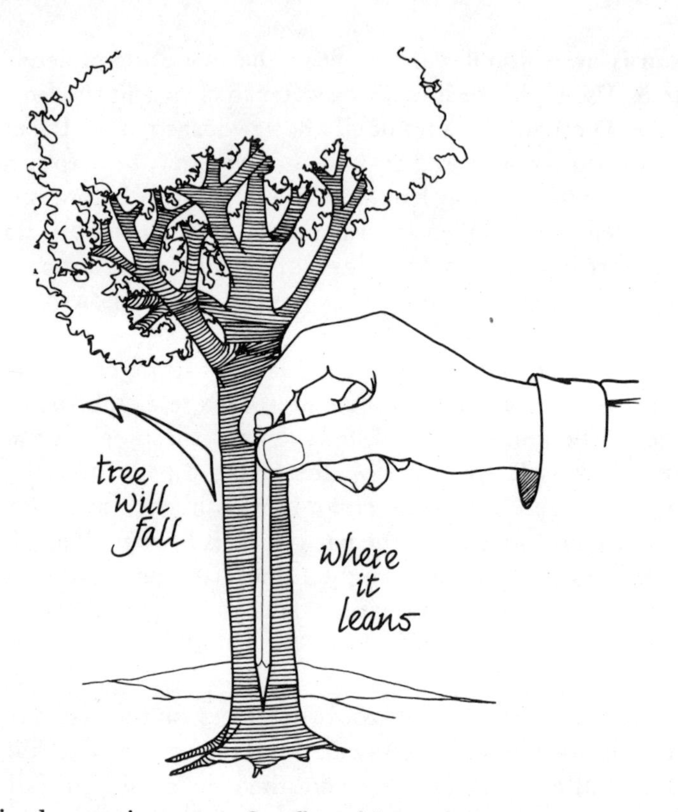

has lived up to its name. So, first thing, plan your escape route. As soon as the tree begins to lean, get going. Always uphill or along the slope if you are working on a hill. A tree can move downhill a lot faster than you can. I prefer to back into standing timber and have two or three trees between me and what's coming down.

If a limb or two appear to overweight the tree in the wrong direction, cut them off. Get up and secure in the tree, use the saw to make a cut about a quarter of the way through on the bottom of the limb, up close to the trunk. Then, lop her off by cutting down to the bottom cut from the top. Without the bottom cut the limb may hang up on a hinge or take a great ragged strip of bark off with it as it goes.

The Felling Cut

To fell, cut a notch in the fall side of the tree. Make it about a third of the way through. First, as close to the base as you can get, cut a horizontal lick. Then start an angling cut about a third

as far above the horizontal cut as the tree is thick. Kick out the notch. Then to fell, go around the back and cut in horizontally a couple of inches above the horizontal portion of the notch. The tree should begin to fall—ever so slowly at first—when you are several inches from the inner face of the notch. The uncut portion will act as a hinge, guiding the tree down as you want it. If

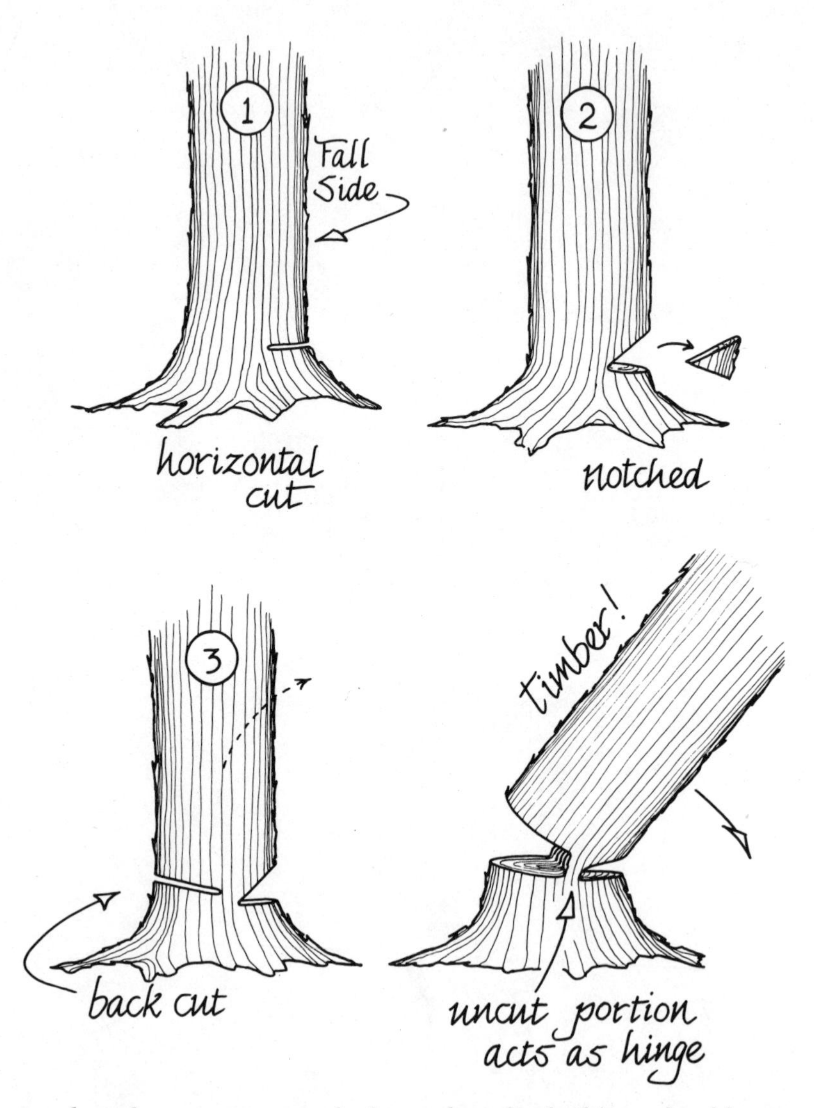

notch and cut were correctly done, though, the hinge should pop as the tree hits the ground.

When you sense that first perceptible movement, turn off the saw and hightail it away. The top can catch in another tree, a hidden irregularity or a hollow in the trunk can affect the fall and the trunk can swivel on the stump and the tree can do you in. I figure I have five seconds to get to safety. I don't know if they hand out awards for the longest distance run in a five-second dash, but if they did, a logger would hold the gold medal. And it's best to practice the dash every time you fell a tree, not on the very few occasions when you suddenly realize you must to stay alive.

The Hung-Up Tree

If the tree settles down on the saw, binding it up, all you have to do is hammer in a wedge behind the bar. They sell plastic ones, but we cut our own on the job when needed. Wedges can also be used to tip a tree slightly if you want to influence its direction of fall. Hammer a good wedge in behind the cutting bar just opposite the direction you want the tree to take or in the cut below a big limb that you suspect might pull the tree in the wrong direction.

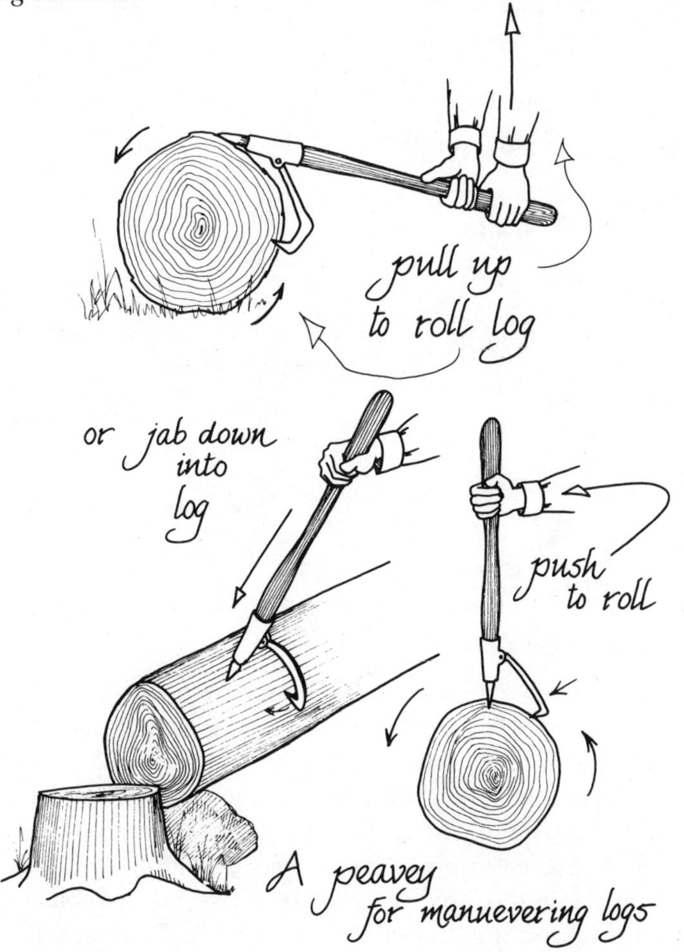

Trees will hang up on you now and again, particularly in thick stands of small cordwood size. Resist the temptation to climb up, joggle it till it's free and then play Paul Bunyan and ride it down. If the butt is completely off the stump, you can use your peavey to try rolling the trunk around. If it is still on the stump, hinge unbroken or even just lodged up against the stump, there can be a tremendous amount of stress at that point. Pull it free and the butt can spring out like a battering ram, and pity the logger that happens to be in its path. It is dangerous to get in under a hung-up tree too. There's no telling when it will fall. I do one of two things. If possible—if there aren't stumps interfering—I loop a length of logging chain around the butt, attach it to the rear hitch of the truck with stout ropes and pull it out. If that isn't possible, a fence stretcher will sometimes get it free. If that doesn't work, we go on to another stand and figure the winter winds and snow will do the job for us. Better, though, to plan each felling well and avoid the problem.

Cutting Up the Felled Tree

Once down, the tree gets another careful survey. The limbs it is resting on will be sprung, containing a great deal of compressed energy. Cut a limb under great pressure and it can break free and spear you. Or, if all the tree's weight is being held by a limb, and you cut it, the trunk can come at you quicker than you can blink. So, be sure you know what is holding up what. And cut into the limb complex carefully. I like to commence at the top, the lightest wood, and trim away main trunk and limbs in short pieces—12 inches for kitchen range wood, 14 to 16 inches for the heating stoves. Trunks stay four-foot-cord size over winter. People who cut wood for a living use the saw to cut four-foot sections, then cut these to length on a cordwood saw with a big, round crosscut blade usually run off an old auto engine. They have log splitters too, and need them. For a one-family fuelwood operation, though, the chain saw is sufficient.

Cutting away a bit at a time, I section the tree as I go, working from the open end. By the time I get to limbs holding the heaviest trunk section, the great mass of the tree has been removed. If a limb does spring, the results won't amount to much because there isn't a great deal of tree attached anymore. If we have room to store them, small sticks are thrown into the truck as cut. Hauled back to the house, they get stacked for the quickest drying, under cover. If not, they just go into a loose pile on the ground. Air circulation won't be as thorough as if they were put in proper ricks but the short length of each stick compensates. By next fall, all but the bottom sticks will be well dried. These ones will go on the bottom of the pile at the house, and by the time we get to them come spring, they'll be dried too.

"Cutting Crib"

The larger wood from trunks or the main limbs of the occasional large tree we take out are cut in four-foot-cord lengths and stacked to dry over winter in the woods. Come the following fall, as we cut new wood for the next year, the year-old sticks are cut to length and split if much over six inches through. The easiest way to cut is to sink four or six posts into the ground in a rectangular crib three-and-a-half-feet long and as wide or deep as the cutting bar of your chain saw. (Six for cutting twelve-inch wood, four for two-foot lengths). Wood is stacked in the crib and you just go down the stack, cutting several logs at a time.

Splitting

To split smallish logs, you can put one up on a stump, or hold it on the ground at an angle, resting against another log as in the photo, and clobber it with a splitting maul. (Wear stout shoes and safety goggles.) The maul is a variety of ax with a heavy, blunt and thick-edged head. Where a much thinner and lighter axhead would fail to split many sticks and lodge in the cut, the maul makes a wide, splitting cut.

For really large logs, you'll need splitting wedges too—like broad, thick axheads without handles. The first cut is made with a maul, you put a wedge or wedges in and hammer them home with a sledge. Generally, we split only very large logs in the woods, leaving most of it for winter exercise.

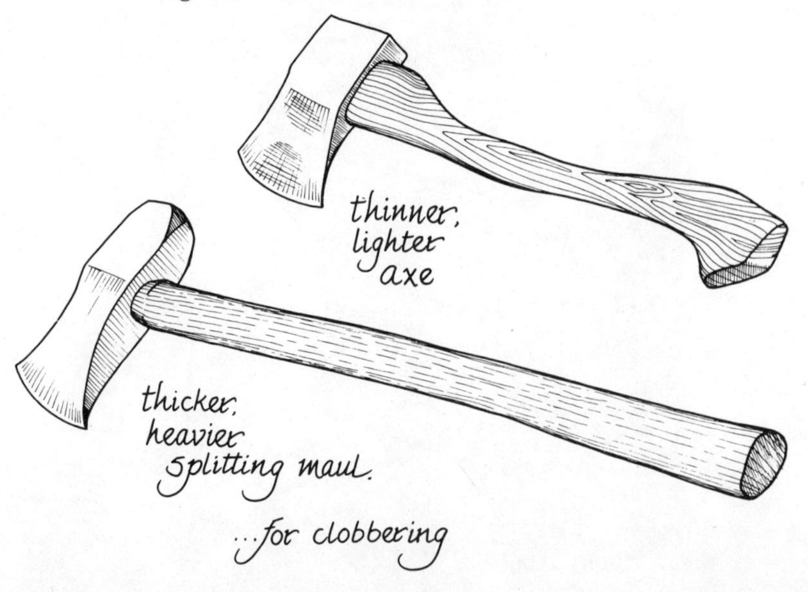

Splitting logs for furnace

steel wedge

Here's how to split wood safely. Hold the stick on the ground with your best foot (the right, if you are right-handed). Have the splitting end—the upper third at least—resting on a half-split placed on the ground in front of your foot's position. Swing the maul (although I'm using the ax here) so it is aimed to imbed in the ground log's forward curve—the leading edge—not in the top or back side. That way the blade can never get to your foot. If the blade slips out or twists off a thin side strip, nothing will get to you. Still, wear heavy boots.

At the end of a stroke, the maul should be in this position: well imbedded in the ground log, but far enough into the cutting log so that it pops in half. It takes some experience to know where to place the cutting log for best splitting. But be sure to move logs around rather than altering the stroke of the maul or ax. You don't want to lose a leg.

With large chunks or knotty pieces, you often need to use a wedge. Here, I'm using the axhead as a wedge to split a knotty piece of white birch. The flat head of the maul acts as a sledge and the chunk will split in three or four whacks.

Here's how to split kindling. Use a good-sized hand ax. Hold a chunk of good-splitting wood, such as white birch, on a log and bring the ax down just hard enough to split out your sticks, pulling the holding hand away just before the ax hits. Too hard a whack and the stick may fly. Too soft and the log won't split and the ax will lodge in the wood. No need to elaborate on what might happen if you don't pull away the holding hand in time.

Cleanup

If you've ever been in the logging country of Maine or the Northwest you've seen the mess a conventional logging operation makes of the woods—tractor ruts everywhere and slash, all but the main log of each tree, lying wherever it fell. But I like to leave our woods as orderly as they were before we went logging. We cut up all branches that are much over an inch thick—the little ones make great kindling or biscuit wood for the range. But there is still a lot of little stuff remaining. In cuttings located close to the road, I haul our big shredder/grinder out from the truck and shred the slash, then haul it home for garden mulch. Far back in the woods we pile it, more or less where the crown of each tree falls, as we cut through the wood. There's no particular reason for doing it, I guess, except out of respect for the woods and the wildlife that will find winter shelter in the piles. A series of little brush piles looks tidier than a general mess. If the trees and rabbits don't care, we do.

You may not agree that a tree deserves respect. But the Indians thought so. They populated every one with a spirit and apologized for cutting it down. Louise and I don't go quite that far, but we can't put out of mind the fact that every tree amounts to so much and is the product of so much of nature's time and energy. Especially the real old-timers. Take that big rogue white pine I mentioned earlier. Why, it must be 250 years old. The trunk is a good five feet through at the base and then some. If trees could think, it certainly would remember the Nipmunk tribe and their campground across the east branch of the Swift (you can see the location from the tree's upper branches), or the Volunteers that set out from Boston in the early 1700s to massacre those gentle people and establish Volunteersville. Or Daniel Shays and his farmer's rebellion that ended in a massacre of another sort when the army regulars hunted them down in the woods just a few miles away. Or the peaceful years when its trunk was surrounded by "mowin' fields" back when Nichewaug was prosperous and John Carter's shutter mill buzzed away just downstream. The old tree has outlasted at least two barbed wire fences that we know of. The rusted ends of the wires twist out two- and three-inches deep into its bark, one behind the other, at the same level on the downhill side. It has Nipmunk arrowheads buried deep in its heartwood too, I'm as sure of that as I can be of anything. And since it's the right distance from the road (in sight but out of hearing) it likely hides a few overgrown heart-and-arrow-and-initial carvings from turn-of-the-century wanderers away from the hayride. It weathered the hurricane of 1938 without a scar. That was the storm that levelled half the county and split one of our roadside maples so it's shaped like a giant slingshot. The tree even survived Clarence Hunt. Its big north-growing limb is where the deer hung that he sold (quarters, halves, whole with hide on or off) before they put in a hunting season and bag limits, and for some time afterwards. And then in his later years, that's where he'd nail up the stuffed head of a 12-point buck on opening day of deer week, then sit in the shade, rocking and waiting, to whoop uproariously when a city dude would saunter by on the road, steadily peering into the woods, then do a double take, fumble at his safety and let go with both barrels into the tree trunk. Hell, it even survived me back when we farmed not far distant. When I learned the hard way that you can't tie up a newly electric fence-wise hog by its neck, even for a few hours, it was that tree I tied her to and it was one big hunk of that tree's root I had to hack out to make burying space for 300 pounds of self-strangled, unbled, thus inedible, pork.

I don't advocate any romantic "poems are made by fools like me, but only God can make a tree" nonsense. But, doggone it, that tree has been a living presence through most every day of recent local history. It's outlived it all and doubtless will outlast me. The tree has seniority and we nod formally when we pass on the cart road.

CHAPTER SEVEN
The Fringe Benefits of Wood Heat

Most folks with the good sense to heat with wood will also have a big garden and may even make their own soap. We do both, and the "waste" from the fires, wood ash, is a big help in each chore.

Wood Ash

You'll recall that the ash contains all the minerals that went into the wood, the water and carbohydrates having gone up the flues in the process of combustion. These minerals were brought up from the subsoil by the tree, whose root structure is almost a mirror image of the top growth, going down many feet. The nutrients the roots bring up are the tree's gift to the topsoil, and to you and I who live on the produce grown there. To rebury a tree's final gift to us in some dump is a waste, almost an insult to the plant that has warmed us and cooked dinner. Ashes should be used.

For the Garden

Pure wood ash contains three garden helpers, in varying amounts depending on the wood. But a good hardwood ash will contain something under five percent of the plant food potassium in the form of potassium carbonate. There is also about two percent of the plant food phosphorous in the form of phosphoric acid. The potassium is what makes strong stems; the phosphorus makes for better roots. These two elements together with the other mineral salts in the ash are alkaline in nature and, like lime, will tend to sweeten your soil. Most garden soils are more acid than most food and ornamental plants prefer, so the addition of wood ash improves the soil's pH—the scale used to measure acidity/alkalinity of soils. Most garden plants prefer a

pH of six to seven, which is nearly neutral. You'd be best advised to make a soil test to see where your land fits on the pH scale. It is possible to oversweeten the garden.

However, you'll get 50 to 60 pounds of ash from a cord of wood, and most gardens will benefit by having that amount applied each year to each 500 square feet, a plot a bit more than 20 feet on a side. You'll find that hot ashes sprinkled on the garden in the cool early spring will help keep down the weeds at the time they are most vigorous. If you've problems with slugs or snails, sprinkle a generous layer of ash at the base of affected plants. The slimy little critters don't like to crawl over the dry ashes. Ashes sprinkled liberally over the crowns of root crops—onions, carrots, beets—will prevent egg laying by the females of many species of "root maggot" fly. The maggots are larvae of bugs that gnaw around in the vegetables. The damage they do is often minor in itself, but their tunnels admit disease that can wipe out an entire crop. We never worry about early spring hatches of maggots; if anything they help thin out the carrot rows. Besides, ash must be dry and unleached out to be effective, and the early spring rains make the ash treatment impractical. But once things dry out in early July, ash is kept in the root crop rows. Plants are mature enough that the mild

alkalinity can't harm them and the bugs have to go elsewhere to plant their broods.

We've found that ashes are an effective barrier to all sorts of insect pests. A ditch around a garden section, some four- to six-inches deep and filled with ashes keeps cutworms out in the sod where they can't get to the young tomato seedlings. To date we've had no problems with bark borers in the young pit fruit trees—peach, plum and apricot. These are larvae of insects that lay eggs at the base of the tree, then the little worm crawls up the

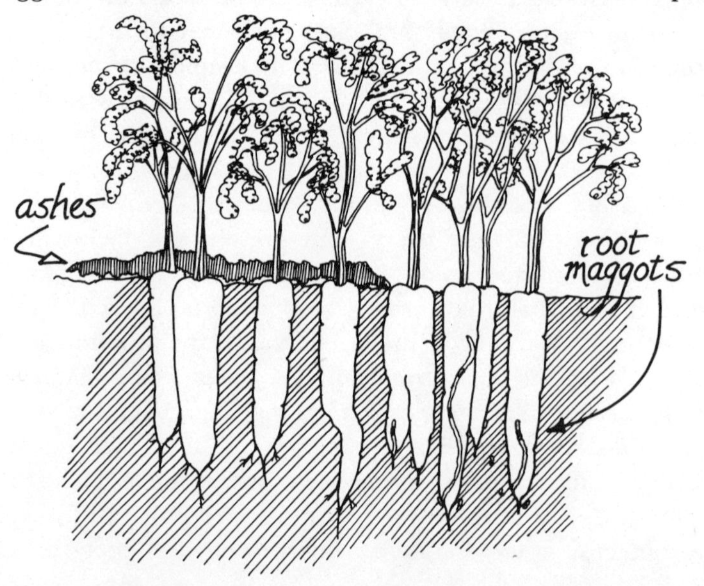

trunk and burrows into a tender branch. A circle of ash is maintained around each tree; not only does it seem to deter the borers, it keeps sod from growing up to the tree's trunk and competing for water and plant food. Dry ash is powdery, slightly astringent and has an unfamiliar smell. I'm convinced that it is an effective deterrent to small, low-to-the-ground garden marauders like rabbits, though I'd not trust it to protect the cabbages against groundhogs or the young corn from racoons.

For Making Soap

The combination of potassium and the other caustic salts in wood ash is most commonly called potash. "Pot ash" is wood lye, which when mixed with animal fats, becomes soap. Grandmaw's lye soap. The early woodcutters turned much of our eastern forests into soapmaking potash for sale to Europe, a criminal waste by today's standards. But we can do the same with ashes left from our own heating fires.

Making Lye . . .

First, you want to be sure to use only pure wood ash. Ashes from newspapers, magazines or other paper waste often contain clay and other additives that you may not want in the wash. You'll want to sift out any unburned chunks, though there should be precious few if you've kept a well-made heating fire. You can make a simple sifter with a square frame covered on one side with hardware cloth and a broomstick for a handle. Just sift out each collection of ashes, and pack the fine ash away till you've a rain barrel full. A big bakery-sized covered flour sifter would work for ashes too, if you have one.

While rummaging around in the old brick ash pit in our cellar, we found a Triumph Ash Sifter. Made around the turn of the century by the Success Manufacturing Co. of Gloucester, Massachusetts, it contains a revolving screen and does a dandy job of sifting ashes, keeping most of it contained so you don't get ash dust all over, as with a hand sifter. I doubt you'll find one for sale anywhere, though it would be easy to make one like it if you've got a knack with tinsnips.

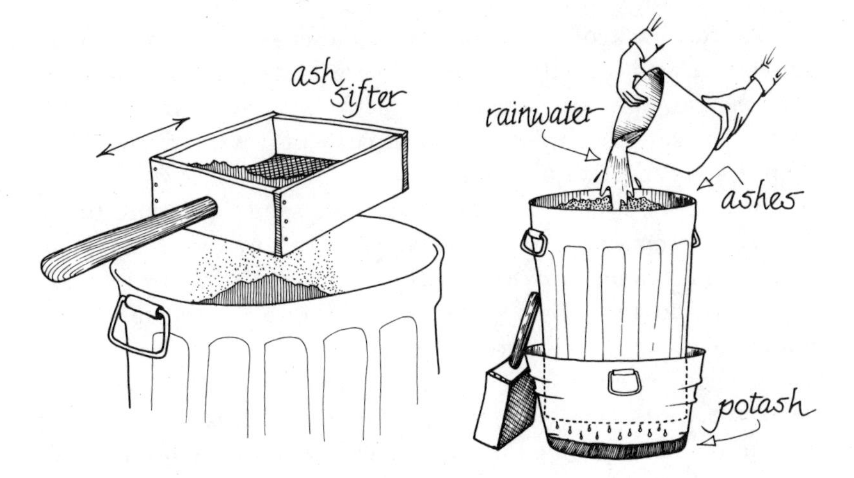

We just set the sifter over a plastic trash barrel with a lot of small holes punched in the bottom and sift until it is full. Then the barrel is set over a washtub and soft water is poured through it—rainwater being the easiest to get. We keep adding sifted ash as it compacts, pour more water on every now and again, then let the leachings evaporate naturally. What's left is potash (lye), and you'll get up to five pounds per barrel. The caked residue in the barrel still has some garden value, though most of the potassium is leached out. So what's left goes on the land. We do make it a point to burn only apple wood for a few weeks now and then. After a particularly careful sifting, leaching and re-washing, the apple wood ash is given a second sifting through a fine screen. Then Louise mixes the ash in water and uses it as a glaze for her stoneware pottery.

Then Soap

When the potash liquid is pretty well evaporated down, we find friends for the kids to spend the day with and make soap. The lye is highly caustic, and children shouldn't be permitted to get anywhere near it. In the morning we fire up the outside fireplace and get a strong cooking fire established. One six-quart pot is filled with the potash. It is boiled slowly till all water is gone. We then dry bake it till all the carbon has burned off and the potash has toasted to a light grey powder. It should be stirred during the smoky, final dry-baking stage. Next we weigh out one-pound portions for each batch of soap.

At the back of the cookstove Louise keeps a one-pound coffee can with an old-fashioned fine-mesh tea strainer over it. Into the can goes all the cooking fat from sausage and bacon, roasts, chicken; any animal fat or mixture of fats will make good soap. Each can, not quite filled to the top, will hold two pounds of fat, and we always seem to accumulate somewhat more than four cans each year. More can be produced by frying up beef tallow. You can buy unsalted lard—expensive, but highly refined and makes lovely soap. On the farm we always saved fat from slaughter of all animals, freezing it till we had enough to justify starting up a wood fire outside under the big old, round-bottomed, cast-iron rendering pot we found in the barn.

In any event, waste fat must be cleaned; for each batch of soap we put two cans of fat into a pot half-full of water, stir and boil for a half hour. The pot is let cool, and you can either pour off the hot fat or let it sit and harden into a cake. The cake can be knocked out and the bottom char plus the top layer with the floating residue can be scraped off.

Now, all we need is a third iron pot big enough to hold fat, water and lye without boiling over, two ladles and a long-handled spoon to stir with. Cast-iron pots are best, though porcelain-lined ware is all right. Never use aluminum pots to make soap. The 12-quart stainless steel canner is of sufficient size to hold our recipe of four pounds of fat plus a pound of potash dissolved in three gallons of water.

Saponification

The soapmaking process is called saponification. I vaguely recall studying it in high school chemistry. How it works I can't recall, but if you combine water and fat or oil with any one of several caustic compounds, it all combines to form soap. Just dump them together in a barrel and you get soft yellow soap that will eat holes in the wood in time and will do likewise to your clothes or the family dog that gets washed in it. But cook the mixture in saltwater and you'll get a nice, white hard soap that is mild enough to use for bathing if it is cured for six months and you've a tough hide. All green soap is caustic at the start and the longer it ages the milder it becomes.

There are several good books out on soapmaking now, and in them you'll find a lot of recipes and good directions. But here

is the old-fashioned three-pot boiled soap we learned to make back before the books were available. First, a pound of potash is stirred slowly into a gallon of cold water in one Dutch oven; this will heat up and can burn, so the emphasis is on slow stirring. The lye water is kept hot while a gallon of water is brought to a boil in the canner and four pounds of fat and a cup of water is heated to the point where the water is bubbling up through the fat. Then as Louise stirs, I dribble in ladles full of lye water and fat till all is mixed.

The glop is let simmer for the afternoon, four or five hours. It should be stirred constantly, but seldom is. Then another gallon of water is brought to a boil, a cup of salt is added and this is slowly added to the mix. All is stirred constantly and cooked slowly till the soap forms on the surface. If you've ever let a bar of soap sit overnight in a soap dish full of water, you recall the ropy, slimy stuff that resulted. That's what new soap looks like. Scoop some out and if the fat and water don't separate, and the mass holds together, you have soap. It is still caustic, and will remain so for several weeks, so it should be handled with rubber gloves. We line a pair of two-foot-square wooden trays that are about four-inches deep with cheesecloth. Then the soap is ladled in and covered to cool slowly. The leftover saltwater is poured into a hole in the ground and buried. Soap will harden slowly from the bottom up. Usually there is a thin layer on top that won't harden completely; it is scraped off and discarded. In about a week the soap will be hard enough to be cut into bars; it is worked out of the trays with the cheesecloth and sliced with a length of fine

piano wire with wooden handles at each end. From the four pounds of fat, one pound of potash and some 20 pints of water we get about 50 standard "personal-sized" bars. They are stacked on their ends back in the trays so air will flow around them freely and they will cure completely, as mentioned earlier, for as much as six months.

Once cured, the soap can be used for about any home cleaning use. Chipping it up into soap flakes on a cabbage grater

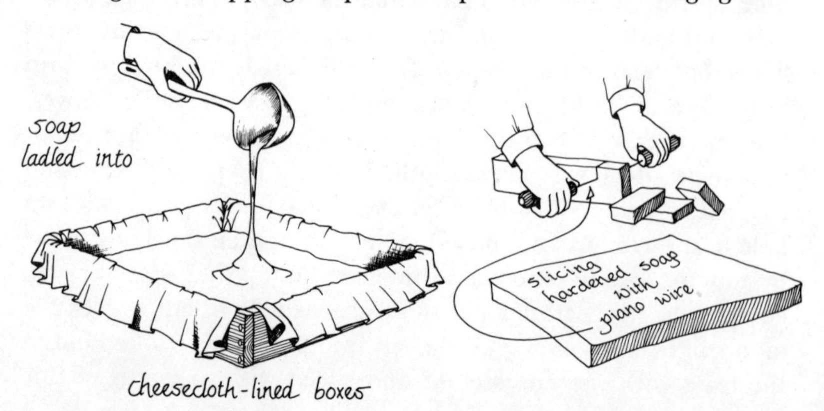

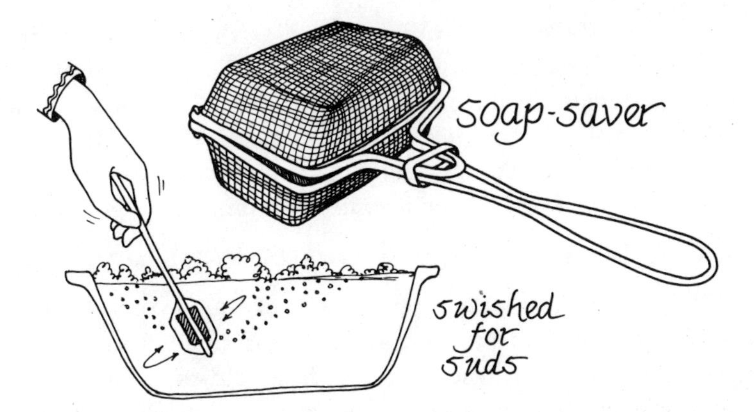

is a lot of work. But for washing dishes we cut bars to fit an old-fashioned soap-saver, a little screen cage on a handle that Grandmaw used to hold scraps of soap. Swished through hot dishwater, it works just fine. For dishes or laundry, you can make a semi-liquid soap. Just chip a bar or two into several chunks and drop into a jar full of hot water. In a few days they will look much as they did when first scooped off the boiling pot. The liquid soap does well with hand washing or the old-style drum and wringer washers. But if you've an automatic washer, don't try homemade soap as modern cleaning technology can't handle the old-fashioned suds from soap you've made for free out of the "wastes" of your wood heater.

Soot

As someone has probably said, if you like the ashes, you'll love the soot (from cleaning the stovepipes, flues and chimney clean outs). It contains many of the minerals found in wood ash plus about two percent nitrogen. Florists use it in potting soil and so should you. We save up soot and add it to the mixture of garden loam, peat moss, sand and lime that we use in the spring to start the tomatoes and other long-season garden plants. Mixed with milled sphagnum moss and varying amounts of loam and lime, depending on the plant's requirements, it is also a good addition to greenhouse or house plant growing medium. It makes any soil a deep, rich black color.

Soot also goes by the names of lampblack and carbon black and is good for any application where a dark coloring is wanted. Masons use it to darken up mortar. It's a good pigment for paint and varnish. Make your own ink by boiling the soot up in a little soapy water. It writes best with a quill pen. For one of those you need the biggest bird you can find; take out a wing or tail feather, whittle the tip off at a slant and split the tip. It takes a lot of dripping and shaking but is a good way to write letters by the light of your wood fire on a winter's evening.

An Old-Fashioned Wood-Heated Wash Day

To get the most washing power out of your soap you'll want lots of piping-hot water. And all the old-fashioned soaps we've made somehow seem to suds up best in an old-time wash tub. If your cookstove has a water front, you are in the scrubbing business. All you need besides soap, your brush, a wash board and lots of elbow grease are a couple of buckets of water, a good fire in the stove, an hour's time and a big dipper. A kettle of water will heat up on the stove top in about the same time.

Our great-grandmothers had a little laundry stove with a small firebox and a wide top. On it went the oval tin "washer" with wood handles at the end in which she heated the wash water. Often lye soap was chipped in and the wash was literally

Washday team

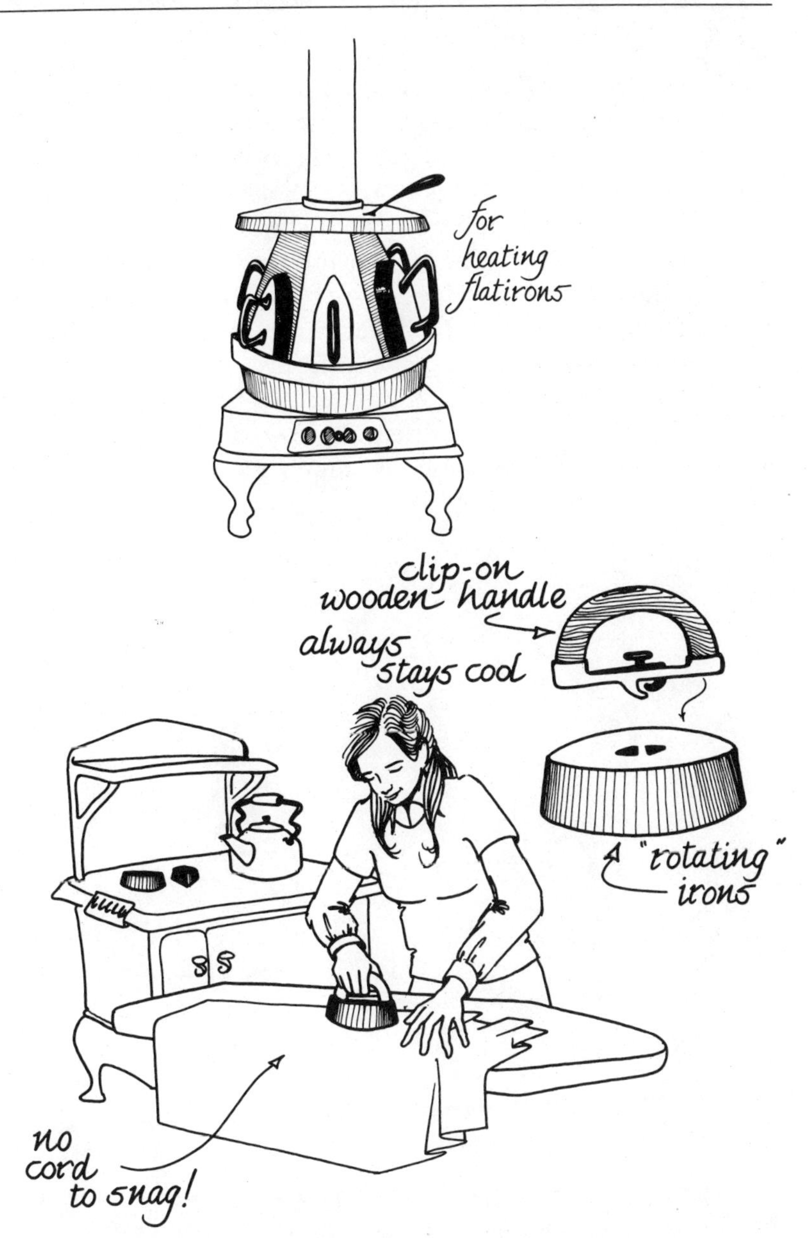

boiled clean. Many old laundry stoves have special holders around the sides for flatirons—different sizes for different chores. You can heat them up on the cook top, too. Grandmother's starched ruffles, lace curtains and other items needing ironing have pretty much gone by the board in today's drip-dry society, but in ironing out your sewing and such, an old flatiron beats any modern, electric, lightweight steam iron. They are heavy, and once hoisted up on the ironing board, do the work for you so long as you keep the bottom waxed (with wax paper). Just keep one or two heating while another is at work. You'll scorch more than one piece of cloth in learning how hot to heat them.

And when it's bath time for the children, out comes the wash tub, and one big kettle of water per bath is heated up. The kitchen becomes the bathroom in winter, of course, The big kitchen range is well fired and with the oven door open, it's the warmest place in the house. And after each small ear has been scrubbed pink, the kids are dried with towels that were warmed up on the stove's racks—and the towels go right back to dry overnight.

Psychological Value of Wood Heat

Finally, don't overlook the intangible benefits of wood heat. First to mind is the simple attractiveness of an open fire, the changing pattern of light in the firebox, reflected to dance on the walls and ceiling. The licks of flame, glow of coals and the infinite variety of shapes and patterns. Bright sounds: snaps and pops and gentle roar of heated air passing into the flue. Then there's the benefit of all the exercise wood heat gives you—forces on you.

I'm sure that in Grandad's boyhood he saw little benefit to the work of hauling in the wood, while his father must have disliked the splitting, and likewise, his mother the cleaning of ashes off every flat surface each week. But they had no choice, and we can't begrudge them all the sigh of relief that accompanied the kerosene stove into the kitchen. But we and our kids (you and yours too, I hope) are returning to the hard work of wood heat willingly, even enthusiastically. I guess we welcome the environmental benefits and freedom from Exxon and its sisters that our willing wood labor brings as much as an earlier generation welcomed the relief from forced labor brought by the kerosene range.

And the exercise you get in splitting and hauling the fuel supply is made somehow more enjoyable by the close contact it gives with the wood. Wood is a bit of nature, a reminder that this log once was a tree in a forest, home to squirrels, birds, perhaps a racoon family. The feel of wood is good. The roughness of the bark, the occasional splinter you have to dig out of a thumb. I don't even grumble (much) when cleaning up the bits of bark, leaves and footprints that always leave a trail after I and the small helpers have brought in the day's wood supply.

Perhaps the prime attraction of the fireplace or stove is the fact of having a central "hot spot" in your home. The romantic symbolism of the fire in the hearth, with its connections to Mom and home cookin', is obvious and best left to more poetic souls than I. But life in a wood-heated home has more variety, more texture and novelty than living in one of your modern split-levels with electric, gas or oil heat. They are probably more uniformly comfortable, but typify to me at least the bland uniformity that seems to be overpowering life these days. With a wood fire, you can go out to the woodshed and split wood in below zero weather, then come back in and toast your snowy mittens to steaming right in front of the blaze. The fact that you can makes the chopping and the cold not only bearable, but downright enjoyable. With a centralized heat source your whole house will exhibit a variety of climates for a variety of activities and people. Near the fire is hot and dry and just right for Grandmother's rocker. The warming shelf above the wood range is grand for rising bread or sleeping cats and the house dog sleeps warm underneath the oven. In the sitting room, facing the fire but over next to the window is where

my reading chair goes; I need the slight chill to keep from dozing and can always pull closer to the heat if need be.

Except for the baby, whose crib goes over the register in the floor of the bedroom just over the kitchen range, people sleep best in cold temperatures. With plenty of blankets, of course. But the chill air coming in from outside carries moisture—good to keep your head and chest from drying out and leading to colds and flu as in many conventionally heated houses. If the cold is too much, pile on more blankets or open the door more or cut another register in the floor.

And, speaking of baby—once he or she puts on a few more years—what's more down a child's line of work than helping bring in the wood? Not that our preschoolers are assigned chores just yet. But a little guy enjoys being able to pitch in with the real work around the place when the mood strikes, and bringing in the logs (he's gotta try the biggest) now and again, gathering up sticks in the woods, and even driving Daddy half nuts by insisting on hanging around to fetch flying sticks when logs are being split, is a real contribution. In a wood-heated house a child—even a very small one—can participate in the real, meaningful task of keeping the whole family warm and well fed. The wood he hauls in is a genuine help and he knows it. It is not make-work or pure play and it's more than several cuts above sitting slack-jawed in front of the TV all afternoon. Which says something good about this (and maybe other) aspects of life at a pretechnology level. Everyone can help. *Really* help, even the smallest members of the family.

And I could go on with wood heat's extra benefits. The cellar is unheated and chilly all year long—fine as is for me if I'm planing down boards for a new table, but when Louise comes down to make a few pots, and the children move into the playroom converted from the old coal bin, it's good to fire up the instant-heating sheet metal stove and warm things as long as needed. Same with the stoves in my little den in the attic, the workshop in the barn and so on. With wood heat you can be just as warm as you want, when you want, where you want. And you find that when you have this flexibility, this variety, you take advantage of it, adjusting heat to activity. The constant change does you good, keeps your head clear and your eyes wide open.

So we can even add a bit of psychological advantage to the physical benefits the exercise of chopping and splitting gives you, the money you can save, the valuable fossil resources you avoid using and the general environmental benefit of burning a clean and self-renewing fuel. Anyone who hasn't changed to wood heat and can, should. We hope you do if you haven't already and that this book helps a little.

Stay warm.

APPENDIX A
Wood Heat's Future

Where wood heat will be in another couple of decades no one can be sure. There will be more folks using it than now though, we can be sure. By some expert estimates, most of the oil will be gone. But we'll have a lot more wood heat options available by then as modern technology comes to bear on such problems as incomplete combustion and creosote buildup. I've heard of several inventive groups that are looking into domestic manufacture of the heavy airtight stoves of Scandinavian design. One, if it gets beyond the design stage, will heat a small house, top and oven cook, heat water for both the domestic washing supply and to heat radiators in distant rooms, all on wood or another fuel of choice.

Several firms among the suppliers listed in the back of the book already make and sell combination wood/gas, wood/kerosene and wood/coal kitchen ranges with water fronts. At least one supplier I know of is working on a combination wood/electric kitchen range for U. S. manufacture, and it seems that every day we see new imports advertised. At this writing, the newest item is the columnar stove, in tall, thin tubular shapes similar to the old railroad stoves. One new import is a lovely little thing, with various shades of brightly colored enamel on iron cast in an almost frilly design, capable of heating a small room and aptly called a boudoir heater. Others come with see-through doors having a Pyrex or old-time mica insert in the door.

And more and more people are experimenting with old ideas and coming up with new ones. Here are a few—novelties now but perhaps a feature of every modern home in another generation.

Some years ago, while traveling in the Orient at Uncle Sam's request, I ran across a few old but good wood heating and cooking ideas that folks might be able to use back here in the petroleum-short future. The Koreans, for example dig a firepit or build a stove at one end of a house and exhaust a large cookstove through a series of tile pipes that run under the floor and connect with a flue at the far side. An underground version of the extended stovepipe. I understand that the ancient Romans did much the same thing to heat up their communal baths. The whole system would have to be fully airtight, and it would take a pretty tall flue to draw heat and smoke under a large house. You'd also want a fireproof floor; the Roman baths were solid ceramic. I would imagine you'd have to do a good bit of experimenting to arrange a series of dampers or other means of adjusting air flow through the system. And how the subfloor tubes are cleaned, I don't know. Brushes on really long handles, I'd guess.

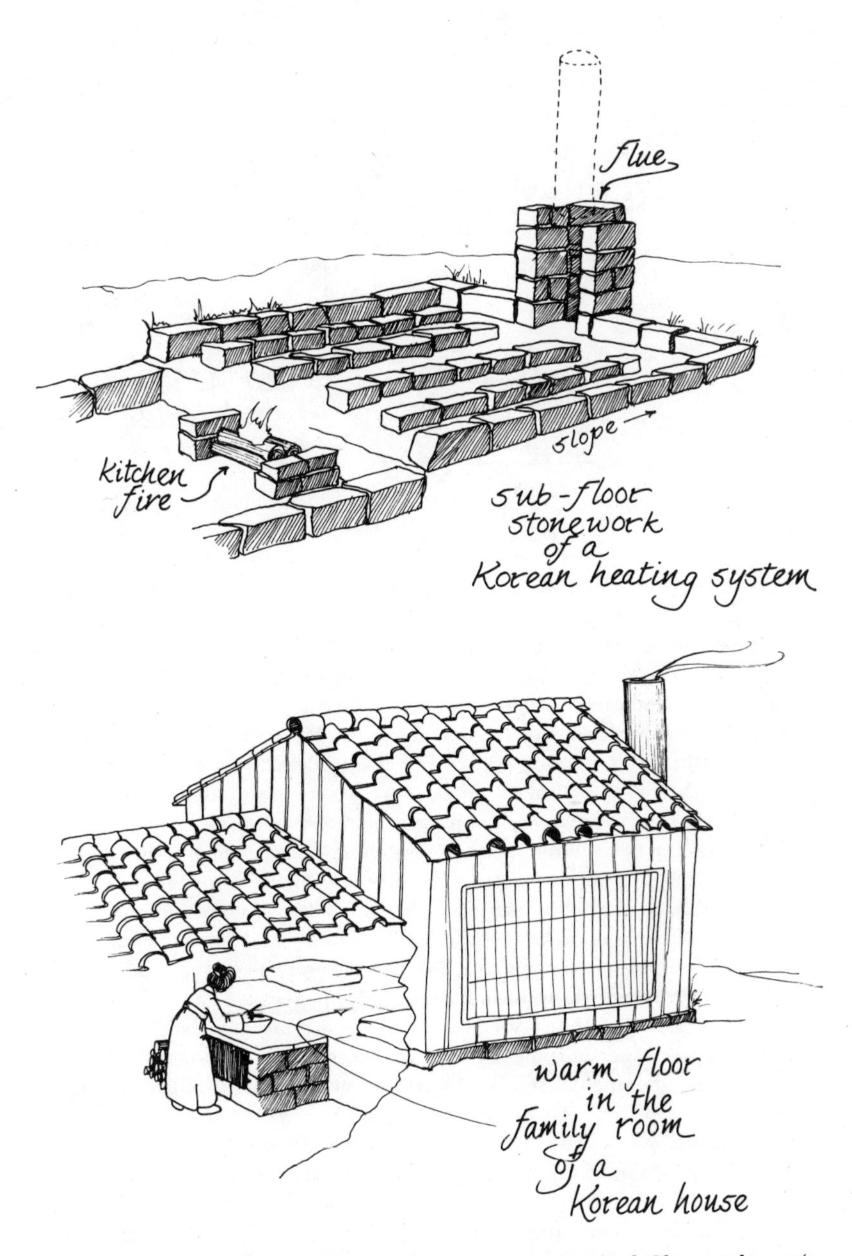

Before their flower culture became an Americanized Chevy culture (or Datsun culture) the Japanese cooked and heated with unvented braziers, shallow pots containing glowing charcoal. The kitchen fire was kept in a series of firepots that resembled built-in hibachis. Rooms were heated with movable firepots, and the most ingenious application I've seen is the pit table. There is a round hole in the floor just knee-to-heel deep. Under the table in the center is

the charcoal brazier. Diners sit on the floor, feet in the pit with a large tablecloth over the table and their laps. The fire warms feet and legs and a warm robe plus hot tea or saki warms the rest.

One warning, though. The houses where this was done were anything but airtight; the carbon monoxide that charcoal produces in great quantity just rose and drifted harmlessly out under the eaves. More than one family in a modern weatherproof home has suffocated as they slept because the fire in the charcoal barbecue that cooked dinner was left to burn itself out inside. Charcoal uses a suprising amount of oxygen, even if you don't see much flame. So don't you dare use it in the Japanese fashion unless your home is thoroughly ventilated. This goes for any other kind of fire too. I recall one of the back-to-the-land publications publishing plans for a sawdust stove that reportedly didn't have to be vented since the sawdust fire can be made almost smokeless. Now, this is fine out in the open, and I've seen such stoves keeping things more or less comfortable for winter work in the mostly open-walled cutting shed of a sawmill. But inside a closed space it would be a killer. Indeed, the more smokeless, and cleaner-burning the fuel, the more dangerous the fire because you won't have the smell of smoke to warn you of danger. Be on the safe side and vent any wood fire. A couple of windows open partway at the top will do for cooking indoors on a brazier (which is what any outdoor charcoal cooker is). I'd never let an unvented fire burn overnight, no matter how many windows are left open.

And the new ideas keep on coming. I read of a young fellow in Maine who hitched up a relatively modern wood furnace to his oil-fired hot-air central furnace. He arranged it so that heated air from the wood burner entered the cold air return of the home heating system and was pumped around the oil burner plenum—the heat exchanger—and into the house. Anytime the wood fire cools down, the oil burner takes over. Smart fellow!

Another development from Maine is being tested by the state's Audubon

Society. The specially designed superinsulated furnace is a true "down-drafter;" the fire practically burns upside down as a baffle system directs smoke back up through the coals. Then the stack gasses are passed through a gravel bed. Most of the heat is absorbed, stored in the gravel to be drawn off by the building's central-heat unit (air or water pumped through the warm gravel just as it is run through a furnace heat exchanger in a conventional central system). This gravel bed also stores heat from hot fluid piped in after being heated in a system of solar (sun heat-collecting) panels. Quite a system and if you'd like to keep track of it and other developments from a state that is one of the last bastions of wood heat in an economy that's pretty well given in to the oil companies, join the Maine Audubon Society, 53 Baxter Blvd., Portland, ME 04101. The membership fee is nominal and their publications well worth the small cost.

And by the time you read this, there will have been further developments. The Maine Audubon system may not be feasible for most people's homes, but it's just a matter of time till someone comes out with a unit combining the best of the old and new ideas that will be priced like any other home heater. And, the home inventors haven't abandoned this technology that we all can understand either. Our wood heat scrapbook of magazine and newspaper clippings on innovative ideas keeps getting thicker. Maybe Louise and I will read about your wood heat brainstorm some day. We'd like to, and once again we truly hope that this book is some help.

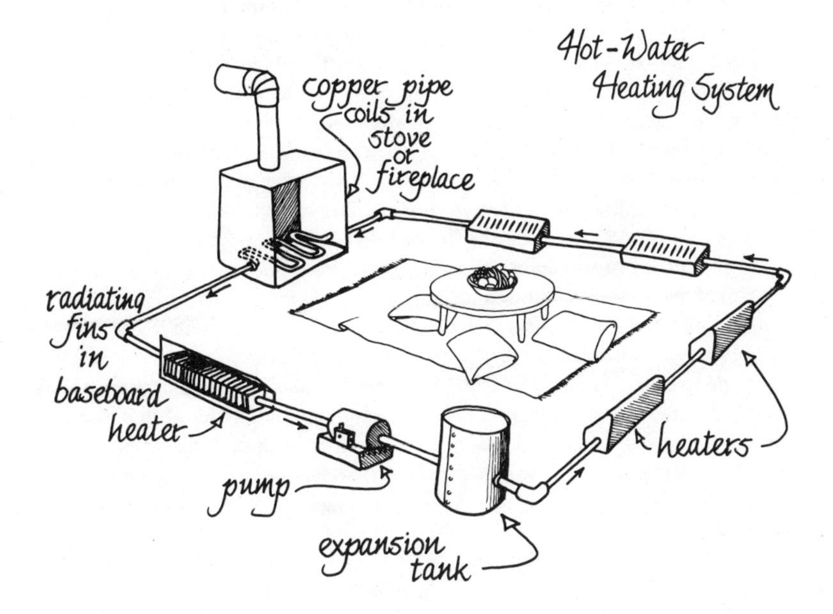

APPENDIX B
Hot Water From the Firebox

Many older wood stoves have ports located at the rear of the firebox that are either used in converting to kerosene burners or to introduce piping to heat water. Our kitchen range has three of these ports and a neighbor has spotted an old, copper hot-water tank in a relative's house. If we haven't got the tank operating in time to get photos for this book, you may be sure it is in (and working—hopefully) as you read this.

Under a gravity water-heating system, you operate on the simple assumption that hot water will rise. The tank is set up on legs so as to be higher than the stove. Water heated in the firebox will rise in the tube and cooler water at the bottom of the tank will replace it. The system would be more efficient if a small pump were attached and perhaps we'll put one on.

Several independent inventors and a few small companies are bringing out devices for producing hot water for washing and cooking and/or for heating the house from a stove or fireplace. One gadget for fireplaces runs tubing halfway up the flue to heat water. Others run coils or a series of S-shaped curves through the fire itself around the back, top or sides of firebox or hearth, or inside or around the outside of stovepipe. Any piping that goes near the fire, once again, should be one continuous piece. The inside of your ash bank can get up to 2,000 degrees F. and that's enough to melt most welding bonds. Don't take a chance.

The piping you ran inside the fire or flue is a heat-exchanger. Attach each end to pipes that run to standard, baseboard finned heaters or braze on your own fins if you have the time and equipment. Then all you need is a small pump to push the water along and an expansion tank somewhere in the system. A possible system is illustrated. Just be sure you get a pump made to move hot water and use sturdy copper or iron pipe—not the plastic hot water pipe; this is made for domestic hot water systems that seldom get hotter than 140 degrees F. The water coming out of the wood fire may be a lot hotter than that. You'll have to obtain pump, expansion tank and valves from a heating contractor, and he can tell you how to put the system together or will do it for you if needed.

Probably the most economical of these hot water units get their heat from the top of the firebox or stove or fireplace or from the flue or stovepipe. This is using heat that would be lost up the chimney otherwise. The main problem of putting any tubing directly in the smoke is that it will eventually become coated with soot and creosote that may burn and surely will reduce its heat absorptive capacity. Better, I'd think, would be to arrange coils around the outside of a

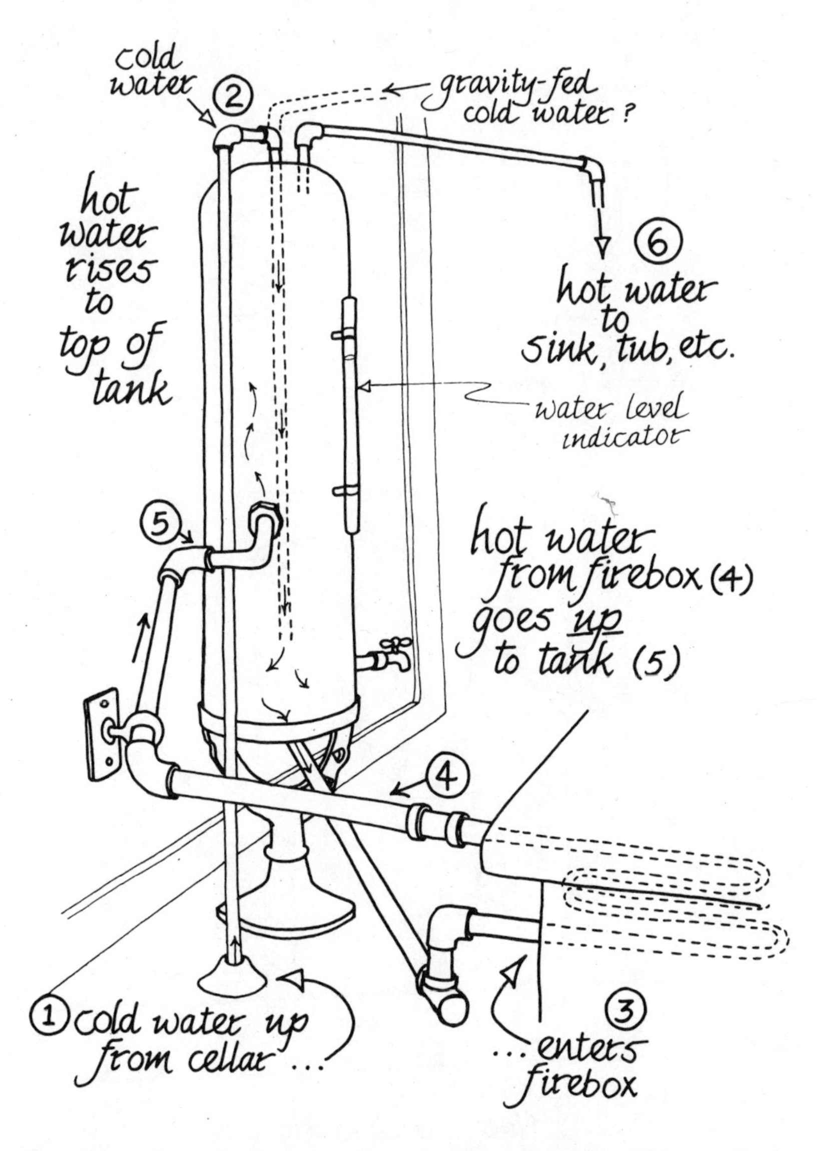

flue liner (or stovepipe as we've already suggested). This would be counterproductive in an installation such as our own living room Combi with an outside flue. Superefficient stoves of any airtight design don't let much heat into the flue and what does get out is needed to maintain draft and keep creosote accumulation at a minimum. There's no point cooling it down with a water heater. However, it would be fine for an inside-house flue. If I ever get around to that log cabin with a Rumford fireplace, I'll build a coil into the flue. I will use round flue liner, which is more efficient than the rectangular, and will mortar a coil of copper tubing around the outside of it, in the space between

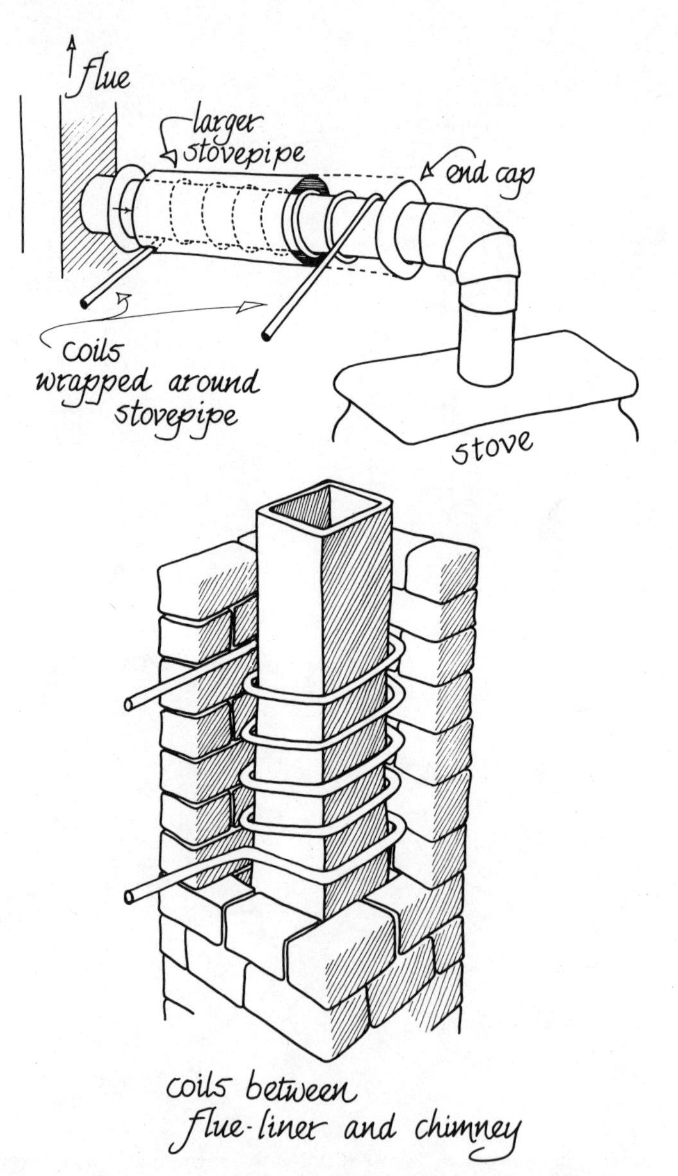

liner and the poured concrete chimney blocks. Then, I'll run pipes in and out through grooves in the blocks, attach a pump and use the system to heat the upper story of the house. To run heat into distant rooms in the lower story, we'll either run pipe under the hearth (through perforated bricks), or I actually will put in the under-hearth, hot-air blower arrangement mentioned earlier. It is conceivable that a fan could be inserted into the chimney, pushing air through the space between liner and masonry, and out into rooms. If you are the first to do it, let us know.

BIBLIOGRAPHY

General Books

Clegg, Peter. *New Low-Cost Sources of Energy for the Home.*
Charlotte, Vt.: Garden Way Publishing Co., 1975.
Coleman, Peter J. *Woodstove Know-how.*
Charlotte, Vt.: Garden Way Publishing, Co., 1974.
Curtis, Will and Jane. *Antique Woodstoves: Artistry in Iron.*
Asheville, Me.: Cobblesmith, 1974.
Gay, Larry. *The Complete Book of Heating With Wood.*
Charlotte, Vt.: Garden Way Publishing Co., 1974.
Glesinger, Egon. *The Coming Age of Wood.*
New York: Simon and Schuster, 1949.
Havens, David. *The Woodburners Handbook.*
Brunswick, Me.: Harpswell Press, 1973.
Home Fires Burning: The History of Domestic Heating and Cooking.
London: Hillary, 1964.
Kauffman, H. J. *Early American Ironware.*
Rutland, Vt.: C. E. Tuttle, 1966.
Mercer, H. C. *The Bible in Iron.*
Doylestown, Pa.: Bucks County Historical Society, 1941.
Smith, Elmer, and Horst, Mel. *Early Ironware.*
Lebanon, Pa.: Applied Arts Pub., 1971.
Stoner, Carol H. *Producing Your Own Power.*
Emmaus, Pa.: Rodale Press, 1974.

Fireplace Books

Lytle, Marie-Jeanne and R. J. *Book of Successful Fireplaces: How to Build, Decorate and Use Them.* 19th ed.
Farmington, Mi.: Structures Publisher, 1971.
Angier, Bradford. *How to Build Your Home in the Woods.*
New York: Hart Publishing Co., 1952.
Sunset Editors. *How to Plan and Build your Fireplace.*
Menlo Park, Ca.: Lane Books, 1973.
Merrilees, Douglas. *Curing Smoky Fireplaces.*
Charlotte, Vt.: Garden Way Publishing, 1973.

Vrest, Orton. *The Forgotten Art of Building a Good Fireplace.*
Dublin, N.H.: Yankee, 1969.
Wigginton, Eliot, ed. *The Foxfire Book.* Vol. 1.
Garden City, N.Y.: Anchor Books, 1972.

Cooking Books

Callahan, Hester. *Cast Iron Cookbook.*
Concord, Ca.: Nitty Gritty Publications, 1969.
Lyon, J. C. *The Fireplace Cookbook.*
Santa Fe, N.M.: The Lightning Tree, 1975.
Wigginton, Eliot, ed. *The Foxfire Book.* Vol. 1.
Garden City, N.Y.: Anchor Books, 1972.

Wood and Logging Books

Allen, Peter H. *Firewood for Heat.*
Brattleboro, Vt.: Stephen Greene Press, 1974.
Brush, Warren D., and Collingwood, Gilt. rev. ed., *Knowing Your Trees.*
Washington, D.C.: The American Forestry Association, 1974.
Forbes, Reginald D. *Woodlands for Profit and Pleasure.*
Washington, D.C.: The American Forestry Association, 1971.
Stephens, Rockwell A. *One Man's Forest.*
Brattleboro, Vt.: Stephen Greene Press, 1974.
Vivian, John. *The Manual of Practical Homesteading.*
Emmaus, Pa.: Rodale Press, 1975.

U.S. Government Publications: from Superintendent of Documents, U.S. Government Printing Office, Washington, DC 20250

Logging Farm Wood Crops, Farmer's Bulletin No. 2090, U.S.D.A.
Managing the Family Forest, Farmer's Bulletin No. 2187, U.S.D.A.
Why T.S.I.: Timber Stand Improvement Increases Profit, U.S. Forest Service PA-901, U.S.D.A.
Wood Fuel Preparations, U.S. Forest Service FPL-090, U.S.D.A.
Public Assistance for Forest Landowners, U.S. Forest Service PA-893, U.S.D.A.
Fireplaces and Chimneys, Farmer's Bulletin No. 1889, U.S.D.A.
Suggestions for the Inexperienced Woodland Buyer, U.S. Forest Service Mimeo Brochure, U.S.D.A.

Safety

From National Fire Protection Association, Int'l.
470 Atlantic Avenue, Boston, MA 02210 ($2 each as of Jan. 1976)

NFPA No. 89m Heat Producing Appliance Clearance, 1971
NFPA No. 10 Extinguishers, Installation, 1973

NFPA No. 25 Rural Water Systems, 1960
NFPA No. 46 Timer: Outside Storage, 1973
NFPA No. 211 Chimneys, Venting System, 1972
NFPA No. 224 Homes Forest Areas, 1972
NFPA No. HS-8 Using Coal and Wood Stoves Safely!, 1974

Magazine Articles

American Heritage
"When Housekeeping became a Science," James Marston Fitch, Aug. 1961
Antiques Magazine
"Stovemakers of Troy, NY," J.D. and J.S. Waite, Jan. 1973
Mother Earth News
"Stovepipe Power," Michael Wassil, No. 24
"How to Make and Use a Sawdust Stove," B.R. Sanbolle, No. 30
"Feedback on Stovepipe Power," Sherman S. Cook, No. 31
"More About Chimney Fires," Paul Stevens, No. 35
"Feedback on the Sawdust Stove," Wayne B. Anderson, No. 35
Organic Gardening and Farming
"Heating Your Home Without Harming Nature," Jeff Cox, Mar. 1973
"Of Wood Cookstoves," Carla Emery, Oct. 1973
"The Woodlot: A Balance in the Ecology of Your Farm," Darrell A. Rolerson, Mar. 1973
"Making the Most of Your Wood Stove," Steve Smyser, Oct. 1975
"The Economics of a Wood Cookstove," Bonnie Fisher, Oct. 1975
"A Fireplace is for Cooking," Jetta Lyon, Jan. 1976
"How to Build a Fireplace Fire," Raymond W. Dyer, Jan. 1976
"Wood Ashes are Worth Money," George and Katy Abraham, Jan. 1976
Yankee Magazine
"Fireplaces were for Cooking," J. Almus Russell, Nov. 1971
"Managing the Small Woodlot," R.M. Bacon, Jan. 1974
"The Shape of Things to Come," C.J. Jordan and J.S. Cole, Jan. 1974

SOURCES

Assorted Stove Models

The Atlanta Stove Works, Inc., Box 5254, Atlanta, GA 30307
Full line of antique and modern cooking range designs, radiating and circulating heaters

The Enterprise Foundry Co., Ltd., Sackville, New Brunswick, Canada E0A3C0
Modern design cooking ranges, box stoves. Franklin fireplaces

HDI Importers, Schoolhouse Farm, Etna, NY 03750
Imports a variety of European designs

King Stove and Range Co., Box 730, Sheffield, AL 35660
Cooking stoves, heaters, laundry stoves

Kristia Associates, Box 1561, Portland ME 04104
Importers of total line of stoves, ranges and fireplace designs from Norway

Maleable Iron Range Co., Beaver Dam, WI 53916
Monarch line of superior wood, wood/oil, etc. ranges, circulating heaters, Franklins

Markade-Winnwood, Box 11382, Knoxville, TN 37919
Good barrel stove kits, fireplace and Shaker-style log burners of contemporary design. Furnace core, in-fireplace stove

The Merry Music Box, 10 McKown St., Boothbay Harbor, ME 04538
Styria made in Austria; heaters and modern-design cooking ranges and accessories. A music box museum too!

Portland Stove Foundry, 57 Kenneber St., Portland, ME 04104
Cast-iron Franklin, parlor, potbelly, Trolla-brand Norwegian stoves. Box, cooking stoves, including antique-style Queen Atlantic. Pipe and accessories

Southport Stoves, Div. Howell Corp., Stratford, CT 06497
Imports the Morso line of European stoves

Washington Stove Works, Box 687, Everett, WA 98206
One of the major manufacturers of cast-iron stoves of all kinds. Accessories including a 70-pound three-legged pot for boiling maple sap, witches brews and missionaries

Weir Stove Co., Taunton, MA 02780
One of the few surviving old-time manufacturers of iron stoves

Wiltons Stove Works, 33 Danbury Rd., Wilton, CT 06807
Several steel stoves of an austere Shaker-type design, a Franklin-type and a dandy cooking range in the works

Circulating Heating Stoves

Ashley Automatic Heater Co., Box 730, Sheffield, AL 35660
Circulators in plain and fancy designs
Atocrat Corp., New Athens, IL 62264
A front-feed automatic stove (cookstoves too)
Brown Stove Works, Inc., Box 490, Cleveland, TN 37311
Still another line of circulators
Foresight Enterprises, Inc., 343 Luinsor St., Ludlow, MA 01056
The N-ER-G Saver stove. Has a cook-top
Locke Stove Co., 114 W. 11th St., Kansas City, MO 64105
Circulators with side-feed
Riteway Mfg. Co., Div. Sarco Corp., Box 6, Harrisonburg, VA 22801
Several models, claims to burn all smoke and gasses

Central Heating Units

Bellway Mfg., Grafton, VT 05146
Custom-built wood, wood and oil, conversions in hot air or hot water furnaces
Dualfuel Products, Inc., Box 514, Simcoe, Ontario, Canada N3Y4L5
Oil/wood combination hot air furnaces
Duo-Matic of Canada, Waterford, Ontario, Canada N0E1Y0
Oil, wood, coal-burning hot air furnaces
Marathon Heater Co., Inc., Box 167, R.D.2, Marathon, NY 13803
A logwood burner capable of 200,000 Btu's
Riteway Mfg., Div. Sarco Corp., Box 6, Harrisonburg, VA 22801
Makes line of wood- and coal-burning hot air and water furnaces (stoves too)

Miscellaneous and Single-Design Stoves

American Stovalator, Rt. 7, Arlington, VT 05250
Another in-fireplace air-circulating stove
Automatic Draft and Stove Co., Lynchburg, VA 24500
Several stove designs, including a sawdust-burner
Cardinal Home Products, 2912 E. 34th St., Cleveland, OH 44115
"Firebird" tubular grate, blower, accessories
Chimney Heat Reclaimer Corp., 53 Railroad Ave., Southington, CT 06489
A stovepipe heat saver and the "carrot top," forced hot-air heating/radiating stove
Fisher Stove Works, 135 Commercial, Springfield, OR 97477
Heater, stove and trash burner in one. Brick-lined steel stove promised to last 200 years!
General Products Corp., Inc., 655 Plains Rd., Milford, CT 06460
Tubular grates and blowers

Hayes-te Equipment Corp., Unionville, CT 06085 and C and D Distributors, Inc., Box 715, Old Saybrook, CT 06475
Sellers of the "Better 'N' Ben's" box stove and accessories—stove vents into existing fireplaces

Isotherimics, Inc., 291 River Rd., Clifton, NJ 07014
Makes heat pipes for heat-saving, in-stove pipe installations

K.N.T., Inc., Box 25, Hayesville, OH 44838
"The Impression," a modern Franklin design with forced-air circulating feature. Comes in black or colored tile

Landau Metal Products Corp., Newman and Ross Inc., 250 5th Ave., New York, NY 10001
"The Kent" stove/fireplace combination

Loeffler Heating Units Mfg. Co., R.D. 1, Box 503-0, Branchville, NJ 07826
Another tubular grate manufacturer

Louisville Tin and Stove Co., Box 1079, Louisville, KY 40201
Stovepipe drum oven (the "Progress," sootless and with thermometer)

Mohawk Industries, Inc., 173 Howland Ave., Adams, MA 01220
A top-loading, downdraft stove of modern design

Patented Mfg. Co., Bedford Rd., Lincoln, MA 01773
"Slip-On Heat Fins" for stovepipes

R and K Incinerator Co., Rt. 4, Decatur, IN 46733
Burn-Easy Lifetime camp stove, a small drum-style stove ready to light

Sandia Indian Industries, Albuquerque, NM 87101
Stove with glass window, water reservoirs for humidification and a design straight out of "Star Trek"

Shenandoah Mfg. Co., Ltd., Box 839, Harrisonburg, VA 22801
A poultry brooder-type stove. Small, but a good heater

Thermal Reclaimation, 939 Chicopee St., Chicopee, MA 01021
Pipe-located heat saver. Thermo-saver

Vermont Techniques, Inc., Box 107, Northfield, VT 05663
A dual-walled air circulating stove to fit into an existing fireplace

Vermont Soapstove Co., Inc., Perkinsville, VT 05151
As of this report (Jan. 1976) the only manufacturer of soapstove griddles. No soapstove wood stoves available at this date; maybe they are making them now. Please send a stamped, self-addressed postcard if you inquire. A small company

Vermont Wood Stove Co., 307 Elm St., Bennington, VT 05201
The "Downdrafter" claims to burn all combustion gasses. They will send you a good little book: "Buyer's Guide to Wood Stoves" which isn't as biased toward their design as it could be

Yankee Wood Stoves, Cross St., Bennington, NH 03442
Drum stoves, assembled and inexpensive

Fireplace-Type Stoves and Equipment

Garden Way Research, Charlotte, VT 05445
Patented grate. By mail only

Greenbriar Products, Box 473, Spring Green, WI 53588
A modern design free-standing fireplace, semi-airtight stove. Accessories for hot-water and hot-air control heating conversion

The Hubbard Creek Trading Co., Star Rt., Box 82, Silverton, OR 97381
Heat Saver brand heat saver

The Majestic Co., Huntington, IN 46750
Another prejob fireplace maker. Dampers, and ventilating brick

National Sewer Pipe, Ltd., Box 1800, Oakville, Ontario, Canada L6J4Z4
Ceramic chimney pots

Radiant Grate, Inc., 31 Morgan Park, Linton, CT 06413
The Dahlquist's grate and cooking accessories

Superior Fireplace Co., 21325 Artesia Ave., Fullerton, CA 92633
Heatforth warm-air circulating fireplaces

Thermograte Enterprises, 51 Lona Lane, St. Paul, MN 55117
Tubular grate

Vega Industries, Inc., Mt. Pleasant, IA 52641
Make a Heatilator fireplace unit and dampers

Vestal Manufacturing Co., Box 420, Sweetwater, TN 37874
Dampers, etc.

Winnwood Industries, 4200 Birmingham Rd., Kansas City, KS 64117
Circulating Franklin-style stove to be inserted in an existing fireplace. Also steel stoves, furnace core

Mail Order Suppliers with Catalogs

L. L. Bean, Freeport, ME 04032
Stove offerings in catalog change frequently. Retail store in Freeport. A major supplier of outdoor gear of all sorts

The Countryside Catalog, 312 Portland Rd., Waterloo, WI 53594
Cast-iron cookware, barrel stove kit, ranges, small but select stove selections including the patented Fisher steel stoves

Cumberland General Store, Rt. 3, Crossville, TN 38555
Most every wood stove design there is, cast-iron ware, accessories such as a 3¾ galvanized wash boiler

Edmund Scientific Co., 555 Edcorp Bldg., Barrington, NJ 08008
Many hobby items including heat savers

W. F. Landers Co., Box 211, Springfield, MA 10010
Perhaps the largest collection of ranges, stoves and accessories

Mother's General Store, Box 506, Flat Rock, NC 28731
Jotul, all kinds of U.S.-made stoves, cooking ranges, accessories, cast-iron cookware and more

NASCO Farm and Ranch Catalog, NASCO, Ft. Atkinson, WI 53538
or NASCO, 1524 Princeton Ave., Modesto, CA 95352
No stoves, but a good mail-order source of logging equipment

Index

D

E

L

M

N

O

P

Q

R

S

T

U

V

W

Y